KT-511-794

WITHDRAWN

Information Generation

Information Generation

How Data Rule Our World

DAVID J. HAND

ONEWORLD
OXFORD

INFORMATION GENERATION

Published by Oneworld Publications 2007

Copyright © David J. Hand, 2007

All rights reserved
Copyright under Berne Convention
A CIP record for this title is available
from the British Library

ISBN-13: 978–1–85168–445–8
ISBN-10: 1–85168–445–X

Typeset by Jayvee, Trivandrum, India
Cover design by Mungo Designs
Printed and bound by Bell & Bain Ltd., Glasgow

Oneworld Publications
185 Banbury Road
Oxford OX2 7AR
England
www.oneworld-publications.com

Learn more about Oneworld. Join our mailing list to
find out about our latest titles and special offers at:

www.oneworld-publications.com/newsletter.htm

To Emily

Contents

CONTENTS

Preface

Our civilization, our quality of life and our standard of living are built on understanding the world around us. Understanding something means we can predict how it will behave, and perhaps even influence and control it. It means we can reduce the uncertainty and doubt which surrounds us. Such understanding and such ability to intervene and control come from facts, information, and observations: they come from data about the world around us.

This book is about such data. It describes what data are, where they come from, and how they are used. It describes the origins of data in prehistory, and the most recent developments in advanced computer programs for analysing and using those data. It describes how data began to shape our view of the universe, and how data are used to shed light on our human condition. It also describes how data can mislead in the hands of the careless or dishonest.

I am immensely lucky. My life is one of adventure. I spend my life working with data. I collaborate with others in deciding how the data should be collected, and then I analyse the data, squeezing drops of understanding and illumination from it. One might even go so far as to say that by such means I have seen things which no-one else has seen.

I owe a tremendous debt of gratitude to many people, far too numerous to mention individually, who have guided, assisted and accompanied me down this road of exploration. They include the many colleagues who share my enthusiasm for this modern voyage of discovery, the many scientists and businessmen and women who have brought me challenging problems, and who have then joined with me in seeking answers, and my graduate students and the collaborators who have made part of the journey with me. Some of our joint discoveries feature as examples in this book.

David J. Hand
Imperial College London
2006

CHAPTER ONE

Let there be Light

'Data! Data! Data!' he cried impatiently. 'I can't make bricks without clay.'
Sherlock Holmes in Sir Arthur Conan Doyle's
The Adventure of the Copper Beeches

The world of data

We live in an age of miracles. We effortlessly travel at hundreds of miles per hour. We fly through the air. We can turn the darkness of night into the light of day just by moving a finger, a power that not so long ago only the deity possessed. Without leaving our armchairs, we can talk to people thousands of miles away. We can see what is happening on the other side of the world. We have even travelled to the moon. All of these powers, and a whole host of others, were the stuff of fantasy to our ancestors not much more than a century ago.

How have we achieved these miracles? What has transformed us from the status of an animal limited by its own muscles, sinews and senses, and given us possession of these almost supernatural abilities?

The answer, of course, is that we have begun to understand nature's secrets. We have learnt something about the vast interlocking structure of cause and effect that is the universe about us. We have grasped why certain things happen, and how to intervene to make other things happen.

In a sense, humanity has been striving towards this understanding since it became fully human. Early man searched for regularities which would permit prediction. Anyone who could predict, such as the shaman or priest who could accurately foresee an eclipse, had awesome power – they appeared to have a direct link to the gods. But the idea of studying such regularities in a formal way grew only gradually. In the dawn of prehistory, just as humanity's caves were illuminated by the occasional fire, so were their thoughts illuminated by the occasional idea of keeping records, of recording rudimentary data by scratching on cave walls. These early thoughts were mere sparks which flickered briefly, lent a moment's illumination, and then died out. Occasionally, however, the sparks combined to form small tongues of flame which danced above the fire, sending shadows across the cave walls of ignorance: carved notches on sticks, or marks on clay tablets. Some of them fell back, while others combined into larger flames of insight and understanding. Gradually these merged: as time marched on, so they came together into great roaring walls of flame. Eventually, they became today's data inferno.

The big shift in all this, lost in the mists of the past and doubtless repeated time after time, came with the change from the subconscious search for patterns to conscious search. And the conscious search for patterns led to the concept of data: the idea of storing the results of observations, of measurements, so that one had a record of what happened, and so that one could pore over the past at leisure. We use data to describe and represent the world, and using it we can search more carefully, more precisely, at greater length, and with less risk than searching the real world.

This process has reached such a pitch that nowadays, of course, we are surrounded by data and subjected to a constant barrage of figures being thrown at us. We open the newspaper and read that the rate of inflation has increased, that house prices are rising, that unemployment is dropping, that school grades are not what they were, that the crime rate has gone up, that the world is gradually becoming warmer, and all of these things have numbers – data – attached to them. At a more mundane level, we look at football scores, we study train and plane timetables, we listen to weather forecasts telling us the expected temperature, wind speed, cloud cover, pollution index, and pollen index, and we study calorie counts or glycaemic index numbers of the foods we eat. At a more exalted level, we see surface pictures of Titan, the moon of Saturn, and are told of the river of data these pictures bring with them: we hear projections of numbers contracting SARS, AIDS, or avian flu, and we study hospital league tables. Not so public, but just as prevalent, are the vast databases that governments and corporations collect and store. To organize a country, to run a corporation, it is necessary to understand its people or customers: it is necessary to have data about them.

These data do not exist in a vacuum. The need for more and more storage capacity, and the need for faster and faster processing of the data have stimulated work on the technological infrastructure that enables today's data flood, that is the development of computer technology. But the pressure is not one way. The development of more powerful computers and database facilities has fed back: if one has the capacity and power, why not use it? Perhaps deeper scientific understanding or greater commercial edge can be gained by collecting even more data and digging into it more deeply. So we have a leapfrogging bootstrap effect. More data requires more capacity, and more capacity permits more data. With competitive drives pushing us, both data quantities and facilities for storing and manipulating those data go on growing and growing.

The power of data

The analysis of data allows us to see things which we cannot see with our raw senses. Just as the telescope and microscope allow us to see things which are invisible to the naked eye, so data analysis allows us to see things which are imperceptible to the unaided senses – not because they are so distant or small, but because they are so subtle or complex. By analogy with a microscope and telescope, one might talk of a datascope. Whereas telescopes and microscopes are made of glass, plastic and metal, a datascope is made of commands telling a computer how to process the data, what numbers to add to what, which to multiply together, and so on.

A table of a million numbers will probably be incomprehensible, a simple mass of digits, but when looked at through an appropriate datascope, structures, patterns and relationships spring out. We can plot the data and note that only a handful of numbers take very large values. We can then correlate the data to see that those cases which take exceptionally high values on one characteristic also take high values on another. This then allows us to summarize the data in simple terms, reducing our million numbers to just a handful which contain the essential properties of the characteristics and their relationships. More advanced and sophisticated datascopes allow us to identify highly complex properties and relationships in the data. We can use these discoveries to predict what will happen in the future, and understand what has happened in the past. We can detect anomalies in the data, noting that sometimes things behave differently, and can then apply other datascopes to tease out when and why this happens. We can also use our understanding of the data to predict what will happen if we intervene in some way – if we prescribe a particular drug, impose a tougher speed limit, or teach students in a different way.

In summary, then, we represent the objects we are interested in by numbers, and then we analyse those numbers to understand the objects. Solar flares are unimaginably distant, huge and hot, and are made of a substance we could not pick up with our bare hands, yet by recording data describing those flares, and analysing it using our datascopes, we can hear the sound of the sun. Atoms are imperceptibly small, and have bizarre counter-intuitive quantum properties, yet by recording data describing those atoms and how they behave, we are able to predict what they will do and even influence them to do what we want. The human mind is difficult even to define, and impossible to see, yet by capturing data about it using brain scans and psychological tests, we can understand what makes it tick – we can probe the mind of a genius. The global economy is an abstract concept, invisible and intangible, yet by measuring international trade, the gross domestic product of different countries and by collecting other data, we can begin to understand how it all fits together.

Collecting and analysing data, and understanding the mechanisms underlying those data are two sides of a single coin. The collection and analysis of data leads to improved understanding of how things work. This improved understanding in turn enables one to design better methods for collecting new data – better measuring instruments, more tightly targeted questionnaires, and so on. The improved new data, more accurate and more pertinent for the phenomenon one wishes to study, in its turn leads to improved understanding of how things work. And so on, in a never-ending data–theory–data dance.

Even at a more basic level, the collection of data about a phenomenon contains within it the seeds for improved future data collection. To collect and record data, we must explicitly state *how* those data are to be recorded, and if we do that, we can think about how the data might be improved. For example, we can see that by not specifying the units in which height should be recorded we leave the door of ambiguity open. Or we can look at a league table of the performance of schools and note that the calculations failed to take account of the qualities of the students being admitted to each school: we can recognize that we may want a table of value added, and not merely of output. In general, if we know how the data were collected, we can adjust, modify and improve the next generation of data.

Data also facilitate communication. If you collect data on a species of barnacle using a ruler which measures in inches, then I know exactly what you are talking about. I can compare the sizes of the barnacles in your patch of sea with the sizes of those in mine. I don't have to transport any barnacles around the world to do this, merely transport, or transmit, the data. We both know exactly what we are talking about, and there is no scope for ambiguity or confusion. Misunderstanding is banished. Of course, we should not claim too much: misunderstanding is only banished if we agree on what we are measuring – if you measure length and I measure height, then confusion will reign.

By representing real world objects in terms of the data describing them, we are *crystallizing* our description of the world. By saying 'this man has a height of 6′3‴' instead of simply 'this is a tall man', we know exactly what we mean. It removes the ambiguity: after all, such a man may not be tall in the context of an Olympic basketball team. This means that we know exactly what we are doing when we analyse the data. Furthermore, data analysis is, to a large extent, objective: it is not influenced by subjective factors such as whether you slept well last night, or anxiety over a forthcoming social encounter. One might expect that a consequence of all this would be that better decisions would be made when they are based on explicit numerical data rather than mere verbal descriptions and, in fact, many investigations of this have been carried out and they show that it is indeed true. For example, it has been shown that methods of medical diagnosis based on numerical measurements are superior to informal medical examinations, that data-driven personnel selection procedures are more effective than job interviews, and that it is better to select credit card or bank loan customers using statistical prediction rules than interviews. At a less exalted level, the eminent psychologist Paul Meehl gave an obvious example of this kind of thing in an article published in 1986. He wrote: 'When you check out at a supermarket, you don't eyeball the heap of purchases and say to the clerk, "Well it looks to me as if it's about $17.00 worth; what do you think?" The clerk adds it up.'

Much of this makes people uncomfortable: as David Boyle puts it in his book *The Tyranny of Numbers* (2000), 'Every time a new set of statistics comes out, I can't help feeling that some of the richness and mystery of life gets extinguished.' I responded to this in 2004 in my book *Measurement Theory and Practice*. I wrote:

> Well, he is right that some of the mystery goes. But my Concise
> Oxford Dictionary defines mystery as 'A secret, hidden, or
> inexplicable matter'. Surely it is all to the good to extinguish such
> things. As to whether the richness goes, on the contrary, the statistics
> (assuming, of course, that they are properly collected and interpreted)
> add to the richness, depth, and understanding of life, deepening our
> appreciation of it, with the potential for making it better.

This unease with formalization, with representing things in terms of numbers which then cannot be argued with, has stimulated anxiety since the dawn of data. It seems to be associated, as in the quotation from Boyle, with the idea that by making things clearer one is removing something of their essence. The truth, of course, is the contrary. By making things clearer, one is approaching more closely to their essence.

Data provide a simplifying representation of the things in which we are interested. A database of employees does not tell us everything there is to know about the employees. It could not: such a thing would be impossible. What it does is contain data on what we believe are the relevant aspects of those employees for the purposes of running the corporation. The data are thus symbols representing the objects we are concerned with, and manipulating symbols is vastly easier than manipulating the objects themselves. I can determine how many women there are in the organization, or how many men aged over 60 merely by counting records in the database. The alternative of walking through the offices counting heads would be much more expensive – and much more likely to yield errors as the employees moved about the building during their work. I can see which of the employees are the most expensive, and I can see where the best savings could be made if we had to downsize, purely from the representation which is given by the data. Data truly provide a lever through which we can move the world.

What are data?

If data are the observations on which we base decisions or draw conclusions, then data have always been with us. Indeed, on this basis, all animals, even very simple single cell forms of animal, use data. An amoeba, for example, will move in the direction of a chemical concentration gradient, although an amoeba does not process its sensory input in any sophisticated way. From this perspective, then, data are the signals, the information from the world which enable us to act appropriately.

Of course, when we speak of data, we typically mean more than the sort of signals to which an amoeba responds. In fact, we typically mean measured and recorded numbers. But just being a number is not sufficient for something to be data: a collection of mere numbers is not data. Such a collection only becomes data when we know something about what the numbers mean; that is, when we establish a link between those numbers and the things they represent – when they are numbers *describing* something. Data are thus a combination of the numbers and what they mean.

Here is a nice little example which illustrates this. Suppose I ask you to calculate the average of 10 and 350. Some elementary arithmetic yields $(10 + 350) \div 2 = 180$. This is perfectly correct as arithmetic, but suppose I now tell you that the numbers represent angles measured in degrees, and ask, again, what is their average? Now, in place of the 180, you may well answer that the average of 10 and 350 is 0.

The word *metadata* has been coined for these descriptive characteristics of data. The metadata tell us the units of measurement, information about how the data were collected, aspects of the relationships between the numbers, what objects the numbers refer to and so on. Accurate metadata are crucial in understanding and enabling the effective use of data: imagine being presented with a database of numbers – a large table in which the rows represented cases and the columns the different characteristics – but without being told what the columns were. Such 'data' would be completely useless: there would be no way to relate it to the real world. It might as well be a table of random numbers for all the good it could do.

A practical illustration of the importance of metadata arises in the European Statistical Office, Eurostat. Eurostat is responsible for integrating data from the National Statistical Offices of the various member countries of the European Union. For example, it has to produce an overall unemployment figure, overall productivity statistics, overall transport figures, etc. Because the various countries calculate their own national values in slightly different ways, all of this would be quite impossible without considerable detail on the formulae underlying the calculations: without the metadata.

Having said all this, if data are more than merely numbers, numbers certainly lie at their core. *Numbers are the basic atoms of data*, and because of their central role, Chapter 2 looks at these atoms in detail.

In fact, the word 'data' has been defined in various ways: factual information, numerical information in a form suitable for processing by computer, values derived from scientific experiments, facts used as a basis for inference, the results of observations or measurements, scientific 'facts', and so on. While differing in details, these definitions convey the essential idea. In particular, we see that data are somehow raw 'facts' – the result of observation and not of analysis, introspection, inference, or calculation. Rather, data are the things which feed into such processes. Theories are what come out of such processes. Theories attempt to explain the data, the raw facts, in terms of overall relationships.

Data can also be seen as evidence. The data items are the clues to how things work. We understand and learn about the underlying mechanism by constructing theories which string the data items together. John Snow's demonstration, in the mid-nineteenth century, that cholera was carried by water contaminated by sewage was based on plotting cholera cases on a map of the Broad Street area of London. Kepler's deduction of the fact that planets followed elliptical orbits was based on Tycho Brahe's painstakingly accurate collection of planetary data. The modern school of evidence-based medicine is an explicit recognition of the importance of having solid data on which to base one's conclusions.

If data and theory lie at the opposite end of a continuum, then *information* lies somewhere in the middle. To be useful, data have to be condensed down, and this condensation transforms it into information. Vast amounts of data are now collected automatically from patients in intensive care units by electronic measuring instruments, but how much information do these signals contain? After all, repeated measurements of blood pressure a few seconds apart are unlikely to show much difference. In an epidemiological study of the incidence of cases of leukaemia around a nuclear power plant, a large amount of data is collected describing the distribution and characteristics of the population which may be at risk, but all this data is condensed down to just a few numbers indicating the risk, and indeed whether there is any elevated risk at all. Information, then, is the *useful content of data*, the extent to which the data permit us to say something helpful about the phenomena we wish to understand.

Data are also descriptive. Each distinct item of data in a corporation's customer database will tell us something about those customers. A database of movies will tell us who directed them, who starred in them, what other movies they have appeared in, and so on. Here the role of the data is not to aid understanding, but rather to enable us to locate any particular fact, any item of detail, we may want.

I noted above that the atoms of data are *usually* numbers. But non-numerical data are becoming increasingly common. Images are an obvious example. In some areas, such as medicine, these have become very important indeed. We have long had photographs of skin lesions, and X-rays of bones, but now we have a wide range of scanning techniques from which we can construct three-dimensional images of what is going on inside us – even in real time so that we can see our internal organs functioning, or parts of our brain light up when we think certain thoughts. Signals provide another kind of non-numerical data. Examples include electrocardiogram traces of how the heart behaves, and electroencephalogram traces of what is going on inside the brain. Speech is, of course, a kind of signal we are all very familiar with.

Although such non-numerical data are reduced to numbers for the purposes of analysis, it is often useful to think of things in the original non-numerical terms. By studying displays of the flight paths of aircraft coming into Heathrow airport, you can easily see where the noisiest areas are likely to be. By studying satellite pictures of clouds, you can see how weather patterns are changing, and what tomorrow is likely to bring.

As far as numerical data are concerned, we use measuring instruments of various kinds to collect data. These could be rulers, thermometers, weighing scales, clocks, hydrometers, hygrometers, ohmmeters, ammeters, and so on, but they can also be more subtle and abstract things, such as the questionnaires used in social survey work, or carefully constructed and validated scales used for measuring pain, depression, or anxiety. It follows from the unlimited number of ways in which data are collected that there is an infinite number of different types of data. In physics alone we have data recorded in units which include the pascal, tor, torr, bar, joule, coulomb, volt, amp, ohm, erg, dyne, watt, mho, daraf, phon, degrees celsius, degrees kelvin, calorie, rad, stilb, nit, lux, phot, bequerel, farad, henry, oersted, tesla, maxwell, weber, bel, roentgen, gray, sievert, curie, and far too many more to list. In psychology, an article in the American Psychological Association's *APA Monitor* in 1992 suggested that some 20,000 tests for measuring psychological, behavioural, and cognitive functioning are developed each year. Not only are we confronted by an avalanche of data, but these data come in an immensely rich variety of kinds, colours and flavours, tapping into different aspects of our lives and of the universe about us.

The cost of data

Data do not simply arrive in one's computer out of the blue. Putting the data there uses energy, costs money and takes time. This is obvious with large-scale data collection exercises, such as space explorations costing hundreds of millions of dollars (and taking years), particle experiments based on billion dollar accelerators, censuses of an entire population, or even smaller-scale laboratory experiments which necessarily involve the cost of the equipment and researcher's time. Even when the data arrive as a secondary activity – for example, in a transaction database stored by a retailer – someone had to write the software that would record the transactions.

In some cases, of course, the resources required to collect extra data may outweigh the information or good which would come from those data. In an example close to my heart, UK universities are subject to appraisal schemes, equal opportunities monitoring processes, a national research assessment exercise, course production audits, research student monitoring systems, student surveys of staff teaching and a national teaching quality assessment exercise. While each of these are doubtless useful as individual exercises, one has to consider the cumulative *measurement burden* that they impose on the organizations. How long might it be before more time is spent on such exercises than on teaching and research? I am sure the reader will be familiar with similar issues in other contexts.

In some contexts, this burden has been recognized, at least in isolated cases. For example, at the time of writing it is possible for a UK citizen to enter the United States without a visa. On the back of the visa waiver form there is a short passage which says: 'Public reporting burden – The burden for this collection is computed as follows: (1) Learning about the form 2 minutes; (2) completing the form 4 minutes; for an estimated average of 6 minutes per response.' It then goes on to say that 'If you have comments regarding the accuracy of this estimate, or suggestions for making this form simpler, you can write to ...' Someone somewhere is clearly aware of the issue and has thought it through.

If collecting data carries a cost, there are also other related issues. Data are often no longer recorded on paper but are instead stored electronically, which has led to immense progress and major advances. But electronic data is by its very nature ephemeral: it can be changed, even deleted and lost forever at the flick of a finger. (Surely everyone has experienced this at some time, when, for example, that carefully crafted email is accidentally caused to vanish into cyberspace.)

Of course, it is not merely electronic data which can be lost. In Chapter 6 I describe the Lanarkshire milk study, an investigation into the effect of milk on the growth of schoolchildren carried out in the UK in the late 1920s. In 1994, when I was trying to get to the bottom of this investigation, I contacted the then Chief Medical Officer of Scotland, Dr R. E. Kendell, to enquire whether any of the original data still existed. He replied:

> We have checked the Scottish Office archives and also asked
> Lanarkshire Health Board to ascertain whether any information

from this study had been kept in their records. Unfortunately, we have drawn a blank on both counts. It was pointed out to us that there had been gross shortages of paper during the War years and that for this reason only absolutely essential documents were retained, others being sent for recycling.

Sometimes this sort of tale has a twist: the data may still exist and the key to its interpretation may be lost, but sometimes cross-referencing it with other data may enable the interpretation to be regained. Charles Babbage gave an example of this in his 1830 book *Reflections on the Decline of Science in England*:

> The thermometers employed by the philosophers who composed the Academia Del Cimento, have been lost; and as they did not use the two fixed points of freezing and boiling water, the results of a great mass of observations have remained useless from our ignorance of the value of a degree on their instrument. M Libri, of Florence, proposed to regain this knowledge by comparing their registers of the temperature of the human body and of that of some warm springs in Tuscany, which have preserved their heat uniform during a century, as well as of other things similarly circumstanced.
> (Babbage, 2004)

A sort of scientific Rosetta Stone.

CHAPTER TWO

The Origins of Data

Measures are more than a creation of society, they create society.
Ken Alder

Numbers, counting, measuring and recording

The concept of number represents one of mankind's major intellectual achievements. It is up there with the invention of the wheel, the discovery of fire, quantum theory and the principle of evolution. At the same time, it is one we all live with in our everyday lives, so we do not recognize what an extraordinarily clever concept it is. In fact, the concept of number is so ingrained within us nowadays that it is difficult even to imagine life without it: it would be like thinking without language. However, the concept of number is not part of the universe; it is one of the ways we represent the world about us, part of the image we make of nature, and not of nature itself.

Psychological experiments have established that humans can only directly perceive numbers up to about four. Beyond that some physical record or mental trick is needed. This mental trick is known as *counting*, and we will explore this in a moment. First, however, let us look at the idea of using physical records to represent numbers. Such records have come in a vast array of different shapes and forms in human history. They represent the first examples of recorded data and permit the one-to-one matching of two sets of objects: the sheep in a flock and the notches on a stick; the days that have passed and the shells on a string; the number of loaves you have sold and marks on a slate. Such simple systems are called tally systems, and many different forms have been used. They are useful because they have a physical continuity: you can put down your container of stones or your knotted thread and know that it will stay the way it was barring malicious intervention or accident, of course, and these, in the form of fraud or error, are still with us in modern data, as we will explore in Chapters 6 and 7.

Some of these physical records are easier to manipulate than others. If you have a notched stick representing the sheep in your flock and you roast one sheep, then the stick has too many notches. A new stick must be carved. On the other hand, with a knotted thread or a container of pebbles it is easy to untie a knot or remove a pebble. Taking this further, if only short-term records are needed, we can keep such a tally using folded fingers. Over a space of moments this is feasible, but it is unreliable for longer periods of time.

Tallies are all very well, but they can easily become unwieldy. Keeping a record of half a dozen sheep in this way is fine, but what about 150 chickens? One would have knotted threads all over the place. The answer is another giant intellectual leap: the idea of groups of things. Instead of 150 separate knots, we could use a large knot to stand for ten chickens – to represent ten small knots. Every time an egg hatched, we would add a small knot, until we had ten small knots, at which time we'd untie all the small ones and replace them by a single large knot. To represent 150, we'd now only need fifteen large knots. And we can take this further. Why not use a super-size knot to represent ten large knots? Then the 150 chickens could be represented by one supersize knot and five large knots. A single thread would be plenty.

Of course, there is no particularly compelling reason for us to use ten in this way, but the fact is that most of the independent inventions of this idea seem to use ten, probably because of the fact that humans have ten fingers. However, other *number bases* have been used. The same idea is used in scoring systems for games and in the British House of Commons library, where four parallel lines are drawn with the fifth crossing them, to clearly indicate a group of five. The Aztecs, Mayans, Celts and Basques used base twenty, presumably including their toes in the matching process. The Yuki of California use eight, probably because there are eight gaps between the fingers.

Such tally systems linger on, simply because they are so very useful. In some cases, they evolve into something more elaborate. The system of Roman numerals has obviously evolved from such a tally system. The first three numbers are represented by simple one-to-one matches: I, II, and III. A group of five (the number of fingers on one hand) is represented by a single symbol, V, presumably a pictographic representation of a hand via the gap between the thumb and fingers. A group of ten is again represented by a single symbol, X. However, rather than representing four by means of the four symbols IIII, just two are used, in IV, 'one less than five'. This apparent simplification has unfortunate implications: it makes life much more complicated when one tries to calculate. (Try multiplying MCCXVI by MMCMXCIX using Roman numerals, for example. Here M means 1000 and C means 100.)

With the move up to distinct objects or marks representing entire groups of items, the concept of number is beginning to come into its own. A new level of abstraction has been reached, in which the group symbol is standing for a numerical entity in its own right. If you like, we have created something which did not exist before: the concept of the number five, or ten, or whatever, independent of and distinct from the things they represent. The number of legs on a sheep or wheels on a car are both represented by the same symbol, and it does not matter what these things are. Some property of the collection, its 'fourness', is being described by the symbol. This really is a key step: we have left the physical world and are now describing it in terms of an abstract representation. We are now *counting*, not simply *matching* objects to tally marks. Moreover, we can manipulate our representation, and the results of our manipulation will map back to things in the real world. If I have a board of pebbles representing loaves, adding pebbles when new loaves are baked and removing them when loaves are sold, I can even start to do speculative manipulation. If I were to bake four more loaves, and two were sold, how many would I have left? Much easier, and cheaper, to juggle with pebbles than bake the loaves. Arrange the pebbles, or beads, on a frame, and we have an abacus. We have now made the first step to calculating.

We can also make things slightly easier if we adopt different symbols for the magnitude of small sets of objects. For example, instead of representing three by three strokes, III, we can use the single symbol 3 (which is in fact clearly derived from three strokes of a marker). This idea leads to the different symbols 1, 2, 3, 4, 5, 6, 7, 8 and 9, the Hindu-Arabic numerals.

Controversy about which was better, an abacus or the Hindu-Arabic system, can be traced back at least as far as the thirteenth century, and lingered on for centuries (probably until the advent of the electronic calculator made the dispute redundant – though I can recall speed competitions between traditional users of the abacus and users of the early electronic calculators). Certainly woodcuts from the early sixteenth century show users of the abacus and Hindu-Arabic numerals in competition.

Representing groups of objects by differently shaped or sized tokens – the larger knots or the differently coloured shells – is one way of indicating a group. Another way, if we are using tokens that have some particular position, such as marks on a slate or knots in a thread, is to let the position rather than the shape or size of the token represent the group it refers to. In particular, every additional place to the left might mean that the symbol represents a group of ten times the size. Of course, to do this you need a place holder, something to indicate that the token is to the left. For example, whereas the symbol '1' represents a single object, the symbol combination '10' represents one object standing for a group of ten objects, and a place holder, '0', indicating that there are no additional single objects: '25' represents two groups of ten objects and five single objects. And so on.

The ground is now set for a revolution. All the elements are in place. We have the concept of natural numbers, and of zero. We have the concept of the position system and of decimal numbers. We have all we need for arithmetic. We have a very powerful tool indeed through which we can represent abstract properties of the world, its 'numerosity', and we can use this tool to derive other properties of the world.

Of course, counting is all very well, but the things it can be applied to are limited: I can count the pencils in my drawer, but I cannot count my height. Only collections of objects can be counted. Height is a continuum – it does not have a unit to be counted in the same way that a collection of pencils does.

This problem is solved by yet another major intellectual breakthrough. It is also a breakthrough that seems so obvious now, we normally do not even notice it is being done. It involves a subconscious alteration in the way we see the world. Instead of seeing height as a continuum, as something which is not a collection, we *imagine* it to be a collection. That is, we define an imaginary object, with an associated height, and we picture our person as being a collection of these objects (or, more accurately, we picture the height aspect of the person as corresponding to the height aspect of the collection of objects). Then all we have to do is count how many objects there are in our collection.

For example, we might imagine a pencil. A pencil has height (or, at least, a length – height is merely a vertical length). We then line up a sequence of pencils, until their end-to-end length matches the height of the person. We call the number of pencils we need the height of the person, in pencils ('George is ten pencils high, whereas Fred is only four pencils high'). Of course, it may not be very accurate – perhaps the total length of the pencils does not match the height of the person very well. We can ease this problem by using a smaller basic unit of length. We could use the width of a pencil as our unit, and then the height of the person would be some large number of pencil widths (George might be 200 pencil-widths high).

By such means, we have transformed the ill-defined question of 'measuring someone's height' into a question of counting.

The *Treviso manuscript*, published in 1478, is the earliest known printed book describing Hindu-Arabic numerals and how use them to do arithmetic. Moreover, whereas other mathematics books had been written with cognoscenti in mind, and had not been aimed at just anyone who wished to learn, in contrast the *Treviso manuscript* was aimed at the 'layman' – it was written in the Venetian dialect, the language of the local masses, rather than Latin, the language of traditional scholarship. Elsewhere in Europe, Roman numerals were still widely used, but the demands of the thriving commercial base of Venice and its surroundings required more sophisticated tools for calculation. We now have the tool, and information about how to use this tool was being broadcast all over the world.

Equipped with the tools for counting, measuring, and recording data, elaborate accounting systems were developed. Such systems allow one to record and monitor a process. They also provide a system allowing one to check inflows and outflows, and, of course, to experiment with alternative scenarios before actually applying them to the real world.

Systems for recording transactions can be traced back as far as one likes: as with all such developments, it is impossible to put one's finger on a date and say 'it started here'. However, they became of central importance as trade became more elaborate and complicated. In his thoughtful and revealing investigation of the role of quantification in Western society, *The Measure of Reality*, Alfred Crosby (1997, p. 201) gives a beautiful example of the complexity of trading some 800 years ago. It involves the merchant Francesco di Marco Datini:

> On 15 November 1394 he transmitted an order for wool to a branch of his company in Mallorca in the Balearic Isles. In May of the following year the sheep were shorn. Storms ensued, and so it was not until midsummer that his agent despatched twenty-nine sacks of wool to Datini, via Peniscola and Barcelona in Catalonia, and thence to Porto Pisa on the coast of Italy. From there the wool travelled to Pisa by boat. Then the wool was divided into thirty-nine bales, of which twenty-one went to a customer in Florence and eighteen to Datini's warehouse in Prato. The eighteen arrived on 14 January 1396. In the next half year his Mallorcan wool was beaten, picked, greased, washed, combed, carded, spun, then woven, dried, teasled and shorn, dyed blue, napped and shorn again, and pressed and folded. These tasks were done by different

groups of workers, the spinning, for instance, by ninety-six women in their homes. At the end of July 1396, two and a half years after Datini had ordered his Mallorcan wool, it was six cloths of about thirty-six yards each and ready for sale. The cloths were dispatched via mule over the Apeninnes to Venice for shipping and sale back in Mallorca. The market there was dull, so they were sent on to Valencia and Barbary. Some sold there, and some were returned to Mallorca for final disposal in 1398, three and a half years after Francesco had ordered the wool.

Crosby draws attention to the care Datini needed to keep track of things, but note also that each step of the above, each exercise involving a task by some other actor, had to be paid for, and, at the end of it all, Datini needed to know that he was going to make a profit: does all that he paid out exceed the amount he would have earned? No wonder there was a need for bookkeeping. Interestingly enough, it was only during Datini's career that Hindu-Arabic numbers began to be used. Prior to 1383 his books write all the numbers out in words.

Although, as the Datini example illustrates, sophisticated systems had been used since before 1400, the name usually associated with the beginning of the *double-entry bookkeeping* system is that of Luca Pacioli (1445–1517). Indeed Pacioli is often described as the 'father of modern accounting'. He was a personal friend of Leonardo da Vinci – they later shared a house in Florence and da Vinci even drew the figures for Pacioli's book *Divina proportione*. The double-entry system simply involves keeping a column of figures for outgoing sums and a separate column of figures for incoming sums, and comparing the two totals at the end to see if a profit or loss has been made. Pacioli gave the earliest known written description of this system in a section of his 600 page book *Summa de Arithmetica, Geometria, Proportioni et Proportionalita*, published in 1494. It has to be said that Pacioli was fortunate in his timing: the development of movable type and the printing press by Johannes Gutenberg (*c.*1394–1468) around 1440 allowed Pacioli's book to be printed and read throughout Europe, and it was translated into several languages.

Although the double-entry bookkeeping system is hardly a major intellectual breakthrough, it has had a huge impact on the way our civilization has grown. Alfred Crosby again (1975, p. 221):

> In the past seven centuries bookkeeping has done more to shape the perceptions of more bright minds than any single innovation in philosophy or science. While a few people pondered the words of René Descartes and Immanuel Kant, millions of others of yeasty and industrious inclination wrote their entries in neat books and then rationalized the world to fit their books.

Bookkeeping, then, gives us eyes, and if we can see what others cannot, then we can lead. Such rigorous accounting procedures formed one of the necessary foundations for the Industrial Revolution, mass production and modern manufacturing and commerce which were to follow in later centuries. Such advances were not built solely on technological progress in the physical sciences. More generally, bookkeeping is the key tool of management. The eyes it provides enable one to see the rocks ahead, and to steer around them. Moreover, management might be at a corporate level, but exactly the same principles apply at national and empire level. Military success hinges just as much on logistic as strategic and tactical principles: an army which runs out of essential supplies such as food or ammunition will not prevail.

Bookkeeping, in the form of accounting, also provides transparency so that others can see that what is claimed to be true is in fact true. Shareholders can study the annual reports to see that the corporation is being run effectively. That is provided, of course, that the annual reports do not conceal or deliberately mislead – recent major accounting scandals involving vast corporations show that this is possible. As we will see in Chapters 6 and 7, the data tell a story, but stories can be fictional.

It is from the detailed records, the extensive and highly accurate data, that models can be built. Actuarial calculation of insurance premiums, for example, hinges on an accurate database of mortality, fires, shipping disasters, or other events. Henry Porter, in an 1853 article on 'the education of an actuary' described statistics (by which he meant numerical data) as being 'the very foundation on which the superstructure of life assurance is raised' though, it has to be admitted, the general feeling amongst actuaries at the time was that numerical data alone were not sufficient, and that a good deal of judgement was also required in order to be able set insurance rates.

In summary, all we are really talking about here is the accurate recording and organization of masses of data – but these data then provide our window to the real world.

Drowning in data

Our modern world is a world of computers. Computers help us to control, manipulate, and understand the world around us, and they do this by collecting data about the world, and reacting to it. However, computers also *store* data for us to manipulate and process. In fact, so much data and so many very large data sets are accumulating that a new technology has grown up to analyse it: *data mining*. This is described in Chapter 5, along with other technologies of data analysis. The anxiety in some quarters is that we are barely keeping pace with the accumulation of data, that we swimming in an ocean of data, trying to keep our heads above the water but all too often failing, and seeing a mass of unexamined data close over us.

One of the consequences of the sheer size of this data avalanche is that we cannot verify the data. We cannot check that it is accurate, true, or correct: we have to take it on trust. Chapter 6 examines this in more detail. Sometimes our trust is misplaced. (How do you know that the information you have just downloaded from the Internet is correct? There are plenty of Internet sites carrying misleading and just plain wrong data.)

With the avalanche of data, numbers such as 'millions' are everyday. Numbers such as billions and even trillions are becoming commonplace. In computer terms, megabyte-sized databases are pretty trivial, and gigabyte or terabyte-sized ones are increasingly common. Petabytes and even exabytes have even been mentioned.

Before we go on, we need to be clear about what these words mean. Things are made easier if we adopt some scientific notation, or perhaps I should have put that more strongly and said that it is possible to talk about these things *only* if we adopt some scientific notation. The alternative is to write several pages of different words to describe a whole zoo of numbers.

The scientific notation we are going to use is very simple: the notation 10^x means the number 10 multiplied by itself x times. Thus, $100 = 10^2$, $1000 = 10^3$, $10,000 = 10^4$, and so on.

Of course, for numbers such as a hundred, a thousand, and perhaps even a million ($= 1,000,000 = 10^6$) we don't really need a simplifying notation. But what about numbers like a billion, sextillion, or nonillion? These are hardly everyday words, and I suspect that most readers do not know what all of them mean, so some way to avoid misunderstanding would be a good idea.

In fact, any reader who was not sure what numbers those words represented has a good excuse: the numbers these words represent are not the same all over the world. For example, a billion in Britain has traditionally meant 1,000,000,000,000 – a million million. In the USA, on the other hand, it has meant 1,000,000,000 – a thousand million. Gradually, the British usage seems to be fading, at least partly as a consequence of the scientific and economic use of these numbers (where else would one have much need for such large numbers but science and economics?). In the following, I will stick to the American usage. So here are some definitions:

- billion 10^9
- trillion 10^{12}
- quadrillion 10^{15}
- quintillion 10^{18}
- sextillion 10^{21}
- septillion 10^{24}
- octillion 10^{27}
- nonillion 10^{30}

There are names for even larger numbers, though, of course, the lack of everyday need for these larger numbers mean that they are normally presented in the 10^x form. If we want to get absurdly large then we can talk of the *googol* (10^{100}) and the *googolplex* ($10^{(10^{100})}$). This last is a 1 followed by 10^{100} zeros. Don't even think about writing it out …. To put the googol and googolplex into context: the number of subatomic particles in the universe has been estimated as being between 10^{72} and 10^{87}.

Now that we have words for very large numbers, let us look at words for very large computer data storage capacities. Development in computer technology has led to huge databases being stored. If we start with some definitions at the bottom end, a *bit* is a 'binary digit': the facility to store something that can be in one of only two possible conditions (e.g. off or on). A byte is a sequence of eight of these. So, a single byte of data has the facility to store any one of $2^8 = 256$ different symbols. This is more than enough to cover all twenty-six letters in the English alphabet as well as the ten numerals $0, 1, 2, …,$ 9. The size of a computer's memory is then measured in terms of how many bytes of data it can store. Not so long ago, one was impressed if a computer could store a megabyte (a million bytes – or, more accurately, 1,048,576 bytes) of data. Nowadays, however, that's nothing. Gigabyte memories (a thousand megabytes – or, in fact, more accurately, 1,024 megabytes) are now commonplace, and even terabyte (a thousand gigabytes – or, more accurately, 1,024 gigabytes) storage facilities are not unheard of. The number 1,024 is, of course, 2^{10} – at base, everything in computer-land works using binary ideas. Beyond that, the word petabyte occasionally crops up nowadays: a petabyte is 2^{50} bytes, or 1,024 terabytes. (Or, if you must, 1,125,899,906,842,624 bytes.) An exabyte is 2^{60} bytes. To put these figures into perspective, it has been suggested that the total amount of printed information in the world is around five exabytes.

These increases in data storage capacity are remarkable. They are described by something called *Moore's Law*. Moore's Law is an empirical observation made by Gordon Moore, a co-founder of Intel. In a paper published in 1965, Moore described how 'the complexity for minimum component costs has increased at a rate of roughly a factor of two per year'. Another variant of this is that the number of transistors in integrated circuits doubles every 18 months or so. Similar claims can be made for data storage capacity. In fact, in terms of hard disc storage capacity per dollar the rate of increase has been even faster over the past decade, and similar claims may be made for RAM storage. Of course, the law is not exact, but application of it over only a few years shows where those gigabytes and terabytes come from.

These extraordinary rates of growth have implications for the computer manufacturing market, which make it a particularly tough sector. For example, it means that hardware 'performance' in the sector increases at about one per cent a week, so that a project which slides by a matter of only a few months will be dramatically behind its competitors. Imagine that translated to other sectors, such as public building works or improvements in service industries!

The overwhelming temptation, of course, is to project these rates of change into the future. Such extrapolation is always risky in anything but the short term, but there seems every reason to expect Moore's law to continue to hold for some years.

Huge modern data sets are arising from a wide variety of sources. Perhaps the ones with which the reader will be most familiar are those that arise in commercial contexts. For example, one of the databases I worked with involved around a billion personal banking transactions. Each transaction involved details of what was bought, where it was bought, how much it cost, and other things as well. Suppose I started examining these transactions by hand, and that I could examine one every second (pretty quick, given the amount of information on each transaction). Since there are thirty-one million seconds in a year, even if I did not stop to sleep, eat, or anything, else, it would still take me about thirty years to examine them all. By which time, of course, the bank would have evolved into something unrecognizable, and they would doubtless not care less about the answer to the question which had set me examining the transactions. My answer would be ancient history.

That example involved around 10^9 records, with the total storage required measured in gigabytes. When we move to the frontiers of science, we hop up orders of magnitude. Particular areas which are rapidly accumulating vast data sets are bioinformatics, astronomy and particle physics.

Microarrays used in biomedical research are structured collections of data measuring the extent to which genes express themselves and interact under varying conditions or experimental treatments. They measure the responses of thousands of pieces of genetic material in parallel, which means on the one hand that complex interactions can be investigated, and on the other that large data sets accumulate. It is common for several thousand responses to be measured in a single experiment, and in some cases arrays have been built which collect information on hundreds of thousands of responses. We are in the realm of gigabytes of data storage. There are collaborative international databases containing information on gene sequences. For example, GenBank® contained nucleotide sequences for over 140,000 organisms, totalling 33.9 billion nucleotide bases in August 2003.

Moving on to astronomy, every day the Hubble Space Telescope transmits ten to fifteen gigabytes of data to astronomers, and the Sloan Digital Sky Survey aims to provide a detailed map of a quarter of the entire sky, listing the positions and brightness of over 100 million astronomical objects, and the distances to over a million galaxies and 100,000 quasars, and which has already accumulated around fifteen terabytes of data. A related project is the GALEX mission, a spacecraft which was launched on 28 April 2003, and which will orbit the Earth for twenty-nine months, collecting image and spectroscopic data of astronomical objects and which will accumulate thirty terabytes of data in that time.

Now let us look at a particle physics experiment. CERN, the European Laboratory for Particle Physics, conducts experiments aimed at probing the very structure of matter. One of these experiments uses the Large Hadron Collider to collide high energy subatomic particles called protons. Each such collision shatters the particles into hundreds of others, and the aim is to see if a particular kind of particle called a B-meson is produced, and, if so, to study its behaviour. In one experiment, forty million groups of protons collide every second, so that, altogether, it yields about forty terabytes of data *every second*. In fact, most of the data is discarded, and some preprocessing at the time the data arrive reduces this unimaginable rate to a 'mere' twenty megabytes per second.

While I was writing this book, just for fun, I occasionally typed the word 'data' into the Google search engine. The number of hits it recorded on 22 February 2004 was 'about 216,000,000'. By 20 January 2005 that had risen to 'about 569,000,000'. And by 8 October 2005 it had risen to 'about 2,140,000,000', although part of the change will be due to changes in the way the engine searched the web.

The numbers we have just been examining are astronomical, so let us try to get some feeling for large numbers by relating them to human scales.

- There are around 7×10^{19} atoms in a grain of sand,
- There are about 2×10^{15} grains of sand in all the world's beaches put together. So Lewis Carroll was right in his poem *The Walrus and the Carpenter* in *Through the Looking Glass*, when he wrote:

'If seven maids with seven mops
Swept for half a year.
Do you suppose,' the walrus said,
'That they could get it clear?'
'I doubt it,' said the Carpenter,
And shed a bitter tear.

- There are around three billion basepairs in the human genome,
- A human has around 30,000 genes,
- The human body has about 10^{14} cells in it,
- There are about 100 billion neurons in the human brain,
- Each neuron in the human brain connects to around 10,000 others,
- A human has 100,000–150,000 hairs on their head (provided they are not bald, of course),
- There are around 6.3 billion people alive on Earth at present (one estimate says about a third of all humans ever born died before history began to be recorded, about a third since

recorded history began, and about a third are still alive; another estimate says that about ten per cent of all humans ever born are alive today. It obviously depends on how you define a 'human being', but either way it indicates a dramatic increase in rate of reproduction over time),

- There are about 10^{15} bacteria living in the human body,
- There are about 10^{18} insects on the earth,
- Around 1.4 million species have been named, and the estimated total number is between two million and 100 million,
- There are estimated to be 10^{47} water molecules on earth,
- Earth is composed of around 10^{50} atoms,
- According to one estimate, our galaxy, the Milky Way, has about forty billion stars in it,
- The Guide Star Catalog II has entries for 998,402,801 distinct astronomical objects,
- There are between 10 billion and 80 billion observable galaxies (though this number continues to increase as telescopes become more powerful),
- There are estimated to be 7×10^{22} stars in the observable universe (can anyone believe that humans are the only intelligent life form in the universe?),
- And, as we noted above, there are estimated to be 10^{72} to 10^{87} fundamental particles in the universe.

Enough!

CHAPTER THREE

A Lever to Move the World

I don't see the logic of rejecting data just because they seem incredible.
Sir Fred Hoyle

Introduction

One of the key defining characteristics of intelligence is that it allows us to think about how the world might work without actually witnessing it. That is, intelligence allows us to construct mental models of the world, and using these we can foresee what might happen, and then take appropriate action. At a simple level, when I see the headlights speeding along the road towards me, I recognize that unless I move fairly rapidly, I will get knocked down, with serious consequences. A less intelligent animal, such as a rabbit or hedgehog, will watch, intrigued, as the glowing discs get larger and larger, until bang! the animal is just a flattened bloody patch on the road.

The key concept here is that of a mental model. Such models free us from the constraints of immediacy, so that we can step back and *think* about what might happen, without having actually seen it happen. Humans, of course, are more intelligent than most other animals, and we can take this further, building much more elaborate mental models. We can imagine what will happen to our agricultural production in a few months time if we do not plant the seeds now; we can borrow money from the bank to buy a car which will allow us to get to a job further away, knowing that we will then be able to repay the bank; we can switch off the electricity supply before rewiring the socket and so on. All of these things involve complex and highly abstract mental models.

Data allow us to take these models yet further – to build models of accuracy and complexity far greater than those which could be built purely in our brains, and to manipulate those models in highly sophisticated and abstract ways. I can tell that Sally (whom I know to be six feet tall) would be able to look down at Claire (whom I know to be three feet tall) even if I had never seen them together. I can calculate that my factory, employing 1000 people and making 500 different types of objects, will make a profit at the end of the year. Using data describing the path of an asteroid heading towards the Earth, I know that, ten million years from now, it will pass close to us. I can tell this even though I cannot and will not see it happen. Data provide a simplified description of some aspects of the world, and by using that description, we can draw conclusions and make inferences about what will happen. Then, based on those conclusions and inferences, we can make decisions and take actions.

One might say that data are *evidence* about how things work.

Of course, data-based scientific models are not the only kind of model. Nor are they the first kind which humanity explored. Magic, superstition and religion are also attempts to construct such models – to find some sense, order and simplification in the buzzing confusion of the world about us – and I will explore these in more detail below. In general, however, these approaches are informal: they do not count and measure aspects of the world, and they do not then try to explain the resulting data in terms of underlying mechanisms, structures and patterns. Above all, they do not make explicit comparison of predictions with observations: in general, those who pursue these approaches do not test them to see how effective they are. This testing is, as we shall see, the essence of science. The lack of such testing means that other kinds of approach are much more vulnerable to subjectivity, and to disagreements which cannot be settled to everyone's satisfaction. If our models are based solely on our opinions, it is hardly surprising that your model may differ from mine. Science, on the other hand, represents an approach to constructing mental models that attempts to remove this subjectivity: science attempts to progress using data that we can all obtain and understand, and move towards models on which we all agree. Obviously this does not mean that scientists never disagree: science is a process of constant revision and improvement of the models of the way things work. What it means is that, when disagreements do arise, when people hold different beliefs about the way the world works, data are collected to resolve the differences, to see which is right, or if, perhaps, some other explanation is needed. This is one of the main differences between the scientific and other ways of explaining the world about us.

A key issue in all of this is what to count as evidence. Data do not exist in a vacuum. As we saw in Chapter 1, a simple table of numbers is not data. It only becomes data when it describes something, that is, when the numbers have attached meaning. But meaning, of course, only takes place in the context of a mental model. We have something of a chicken and egg situation here: numbers are only data in the context of a model, and a model can only be constructed when it has data to describe. Clarifying this question of what should count as data has been part of the progress towards more objective models and better understanding of the world about us. That 'progress' is the right word is illustrated by the technological world around us: we have cars, planes, computers, microwave ovens, and so on, only because our models provide good and tested descriptions of the world. That is, only because the data we have collected do provide relevant information about how the world works. It was not always obvious that the configuration of tea leaves, or the properties of intestines of sacrificed animals, or the two-dimensional star patterns we see from the Earth, or the obscure sayings of oracles do not provide useful evidence, useful data, about how the world works.

Even in the context of simple numbers collected by counting or measuring, it is not obvious what should count as data. Numbers have always had an associated aesthetic, magical, mystical and religious significance. How do I know that such properties do not reflect real attributes of the external world? Perhaps a clue is given by the fact that different cultures attach different significance to different numbers. If seven is regarded as lucky by some people and unlucky by others, perhaps this suggests that 'luck' is, in fact, not a real property of the number seven. The truth is that the magnitude of seven, and facts which can be deduced from this magnitude, are its only real properties. This is what makes the concept of number so important. A number is interpreted in exactly the same way by everyone. When someone says they have three books, you know exactly what they mean. Numbers are objective; everyone means the same thing by them. We might summarize this by saying that numbers travel: wherever they go they mean the same thing.

Of course, in general, symbolism has always been a key aspect of humanity's attempts to understand the universe. The trick is separating the aspects of these symbols which do describe aspects of the real world from those which are imposed by human beings. In fact, one type of genius is surely the ability to recognize what matters: what can be disregarded as an approximation or irrelevance when building a model, and what should be included as a key component. Newton's appreciation that the brightness of the moon should not feature in his theory of gravitation is an example of this genius.

Magic

There is a persistent myth that the ancients possessed great understanding of how the universe works, but that these secrets have been lost over the course of time as a consequence of wars and natural disasters. This is obviously psychologically appealing: the alternative is that no-one really knows what is going on, and never has! The renaissance rediscovery of Greek writings that had been concealed and protected by Western monasteries during the Middle Ages lent credence to the myth, but the idea is more generally pervasive. Indeed, the very origin of humanity, as told by Genesis, is an example of such an idea: that Mankind has been cast out from perfection in the Garden of Eden. Apart from the great religious books, there are also other examples of ancient works (and some perhaps not so ancient forgeries) which also hint at lost knowledge: for example, in Nostradamus's *Centuries* (published in 1555) Century 1, Quatrain 25 says 'the lost thing is discovered, hidden for many centuries'; the Cabbala, with origins going back to the first century, is supposed to contain secret knowledge; John Dee's *Liber Mysteriorum*, written in 1581–1583, is another example. John Donne's 1611 poem *Anatomie of the World* also illustrates this belief in lost secret knowledge:

> *What Artist now dares boast that he can bring*
> *Heaven hither or constellate any thing,*
> *So as the influence of those starres may bee*
> *Imprison'd in an Hearbe, or Charme, or Tree,*
> *And doe by touch, all which those stars could do?*
> *The art is lost and correspondence too.*

Things are complicated in some cases by a tradition of writing things in a style and language which would be comprehensible by the learned, but not by the common man. There must also have been a strong temptation to cheat, something which is made easier if one wrote in an obscure language or style (Nostradamus's predictions are a good example of this). This lent an air of mystery to the material: it suggested that if only one could find the key then great secrets would be revealed.

Someone who had read and understood the ancient texts, then, was thought to be in possession of ancient and previously lost secrets. Such people, such magicians, had no supernatural powers beyond those of other men but simply knew things about the world that others did not – and could exploit this knowledge. This is why, during the Middle Ages, machines were associated with the work of magicians. If a machine worked by some means which was obscure, then it was magical.

Of course, as with modern science, there were binding threads which held magical traditions, theories and beliefs together. One is the idea that behind magic lay hidden or *occult* causes and interactions. The word 'occult' simply means 'beyond the range of ordinary knowledge', so that a magician was merely someone with a superior knowledge of these hidden interactions. In fact, magic and science are similar in this regard, because modern science is full of 'hidden influences'. In physics, think of gravity: you cannot touch, feel, taste, see, or hear it, and yet it determines your entire life, from stopping

you floating off into space to being responsible for the very condensation of the sun and planets from drifting interstellar dust. In
economics, think of Adam Smith's 'invisible hand' steering the
economy. There are any number of similar examples. One might
think of modern science as being concerned with postulating hidden
causes that lead to simpler explanations for what is going on in the
world. From that perspective, magic and science are not so different. In fact, of course, there is more to it than this, and later sections
of this chapter will explore this matter in more detail: science does
not merely postulate hidden forces, but requires them to pass some
pretty stringent tests. We are back to this word 'test' again. Without
passing these tests, anything could be proposed.

Another binding thread which held much of magic together was
the *doctrine of signatures*. In the Middle Ages, and indeed for hundreds
of years later, the spiritual world was just as real as the physical world
about us (just think of the effort and resources which went into
building the world's great churches and cathedrals, and the psychological commitment that this represented), and this spiritual
influence was a formative influence on the ideas and thoughts of
those trying to make sense of the world. Even Isaac Newton was a
man of his time, and was certainly not immune from this – his life
was dominated by religious and mystical influences. Within this
context, the doctrine of signatures essentially maintains that,
because God created plants with humanity in mind, he also gave the
plants signs which indicated how they should be used for the benefit
or otherwise of mankind. Because we eat plants, and some are mani-
festly bad for us (tasting foul or making us vomit, for example), it is
perhaps not surprising that a complex theory of the relationship
between plants and humans grew up. Versions of this doctrine seem
to be widespread in human civilizations, stretching from east to
west. In ancient China, for example, yellow and sweet plants were
related to the spleen, red and bitter to the heart, green and sour to
the liver, and black and salty to the lungs. In the West, sometimes the
signs are obvious: maidenhair fern was regarded as a cure for bald-
ness, mandrake was regarded as an aphrodisiac in females, walnuts

were related to the brain (which they resemble), plants which lived a long time were supposed to promote longevity in humans, plants with yellow sap were good for jaundice; and so on. '*As above, so below; as within, so without*', as the Hermetic philosophers put it.

The doctrine of signatures is ancient, and doubtless one which was reinvented time and again over the course of human history. The great philosopher and physician Galen (AD 131–200) alluded to these ideas, but the real origin is generally attributed to a German shoemaker, Jacob Boehme (1525–1624), in his books *Aurora* and *Signatura Rerum*, which appeared in the first half of the seventeenth century. However, the name most consistently attached to the doctrine is that of Paracelsus. Professor of Medicine at the University of Basel, and originally named Philippus Aureolus Theophrastus Bombastus von Hohenheim, Paracelsus travelled widely, treating people according to this theory, and is a key figure in the origins of modern chemistry.

But is the doctrine of signatures so different from a scientific theory? It is a proposed explanation of the way things work, so perhaps we should accord it the same status and respect as, say, the atomic theory, relativity, evolution, or quantum theory. Furthermore, like those modern theories, it can be applied in practice. Unfortunately, it turns out that a crucial aspect is missing: the doctrine of signatures does not satisfy the rigorous tests which have come to characterize science. In general, predictions based on the relationships it suggests do not hold: its predictions do not match the data. Put another way: *it simply does not work.*

Yet a third important binding thread in the West was the medieval cosmological notion that described everything on earth as being determined by heavenly events, so that planetary qualities were reflected in earthly objects. Once again, though superficially appealing, and making a certain immediate sense (especially in the context of the spiritual world of the time), this theory does not hold up: its predictions fail. Astrological forecasts do not match observational data, though this is often concealed by presenting the predictions in an oracular manner, so that there is freedom to interpret them in a way which appears to match whatever did in fact happen.

A fourth important magical theme, and one which also seems fairly universal to humanity though with different manifestations, is the mystical role of numbers in influencing human life. I have already referred to the fact that people often attach non-numerical characteristics to numbers (ever been asked what is your lucky number?). Pythagoras believed that numbers had mystical significance, based on the universal way they represented aspects of the world about us. So some numbers are regarded as intrinsically lucky and others as intrinsically unlucky. In the West, for example, thirteen is characteristically regarded as unlucky (many hotels do not have floors labelled as thirteenth) because it was the number of people that attended the last supper. Thirteen is also the number of witches supposed to be present at a witches' sabbath. Of course, as I noted above, the lack of substance to these supposed properties of numbers is indicated by the fact that different cultures ascribe different properties to them. The truth is that their magnitude is their only property, and magnitude is universal across all cultures.

There is a strong link between numbers and cryptography, so that numbers can contain hidden meaning, and this has strengthened the imagined relationship with magic. Numerologists, in particular, have ways of translating names, in letters, to numbers, and then interpreting the significance of the resulting number. In the early sixteenth century, partly for this reason, mathematics was regarded with suspicion, and considered related to magic.

In the modern world, magic and science are typically seen as in opposition: magic is 'supernatural', whereas science is 'natural'. However, this is not really accurate. The truth is that magic was a precursor, an ancestor, of science. Indeed, one might even argue that magic evolved into science. A magician was someone who, it was believed, knew how nature worked, someone who could predict what would happen, who could say what would happen *if* .... It was by virtue of this unusual knowledge that a magician could do things that defied the understanding of ordinary men. Since there was so much about the world which defied the understanding of ordinary men, a magician was indeed a powerful and dangerous figure.

However, much of that 'knowledge' was not based on sound practical evidence (on data) and as men began to strip away those beliefs
which had no basis in fact, to jettison the relationships which had
been invented rather than discovered, so the domain of magic began
to shrink. More and more began to fall into the realm of science:
something which followed from understood, *testable* and *replicable*
laws of nature, in which the observations and predictions matched
the data. As this has continued to the present day, so the realm of
magic has continued to shrink. Or, from another perspective, with
the growth in understanding of the physical world which began to
flourish in the seventeenth century, magic evolved into science. One
might even say that scientists are today's magicians.

This last sentence is not so far-fetched. As the opening paragraph
of this book illustrated, today's scientists or, more generally, anyone
using the technological products of today's science, can do things
which would not long ago have been regarded as miraculous, certainly as magical. This recognition, that early magic was a technology
(albeit a poor and unreliable one) based on supposed relationships,
brings new meaning to the observation by the British science and
science fiction writer Arthur C. Clarke (the inventor of geostation-
ary satellites) who wrote, in *Profiles of the Future* (1962), that 'Any
sufficiently advanced technology is indistinguishable from magic.'
Of course, Clarke was not the first to make such a statement. Roger
Bacon (*c.*1220–1292) had said some 700 years earlier that 'many
secrets of art and nature are thought by the unlearned to be magical'
and some 400 years ago Martin del Rio (*Disquisitiones Magicae*, 1608)
described magic as 'an art or technique which by using the power
in creation rather than a supernatural power produces various things
of a marvellous and unusual kind, the reason for which escapes the
senses and ordinary comprehension'. The case of Roger Bacon is
particularly ironic because history twisted his reputation and represented him as a sorcerer ('Doctor Mirabilis').

One of the central tenets of science, one might even say a part of
the definition of what science is, is that it is about gaining *understanding*. Secondary to this is the application of this understanding to

achieve some end. This application is *technology*. One difference in emphasis between magic and science is that the aim of magic was not understanding for its own sake, but rather was to use the understanding – which I suppose means that one might describe magic as technology, more than science. This is certainly one way in which Francis Bacon (1561–1626, no relation to Roger Bacon) advanced the cause of science as distinct from magic. For example, in his *Novum Organum* (1620), Aphorism LXX, he says: 'So must we likewise from experience of every kind first endeavour to discover true causes and axioms; *and seek for experiments of Light, not for experiments of Fruit*' (Bacon, 1994) (my italics). He is distinguishing experiments which increase understanding (those which shed light) from those which give us the knowledge to do things (those which yield fruit). Magic was transforming into a new kind of entity.

If magic was aimed at the study and use of hidden forces and influences, physical and otherwise, governing the way the world worked, much of it was based on observation. Unfortunately, simple observation is an unreliable guide, even if it is superior to reliance on ancient tomes. Events, even the seemingly most improbable, do happen by chance, and to make general inferences from them can be risky. Just because a friend was hit by a car shortly after a black cat crossed his path does not mean that the latter is a sign of bad luck. It was simply a coincidence. Superstitions probably have their origins in such observed chance events. We see, once, that walking under a ladder results in a painful head wound from a dropped object, and make the general inference that it is unlucky. In an analogous way, sometimes people would recover from an illness in the natural course of things, whether or not they took a medicine. But the patients of a physician prescribing a remedy for the common cold might easily be persuaded that it is efficacious, since they always eventually got better after taking it. I am reminded of the aphorism that a doctor's job is to keep the patient happy while nature takes its course.

What is clearly missing from these situations is some way of separating the effect of genuine causes from chance. Without this, we are in danger of making the mistake Francis Bacon cautioned us against: 'God forbid that we should give out a dream of our own imagination for a pattern of the world: rather may He graciously grant us to write an apocalypse or true vision of the footsteps of the Creator imprinted on His creatures' (from Bacon's 'Plan of Work' for his *Great Instauration*). The fact is that 'giving out a dream of our own imagination', is exactly what is done in superstitions and religion, which do not rely on clearly established causal influences.

How can such influences be discovered? The answer lies in the notion of the *experiment*, the *deliberate testing* of one's theories to see if they hold up. Experimentation represents a step beyond mere observation because one is deliberately contriving situations and controlling other possible influences. It represents a step beyond mere contemplation because one is actually looking at data to see what happens, rather than merely speculating. And it also represents a step beyond the reliance on ancient authorities because one is looking at what actually happens. As Francis Bacon put it in *The Proficiency and Advancement of Learning* (Book I, 1605): 'The aim of magic is to recall natural philosophy from the vanity of speculations to the importance of experiment (Bacon, 2002)' Science did not yet exist, but Bacon was creating it.

Francis Bacon saw what we now call science as a way of improving the human condition. Whereas people nowadays tend to see science and religion as opposites, Bacon saw them as working together to replace humanity in its position prior to the fall: at the end of *Novum Organum*, he wrote: 'For man by the fall fell at the same time from his state of innocency and from his dominion over creation. Both of these losses however can even in this life be in some part repaired; the former by religion and faith, the latter by arts and sciences.' He saw science and religion as working hand in hand.

In summary, modern science did not develop as an alternative to traditional magic; rather, it developed as a natural extension of it.

Science: the genesis

One can define science in various ways. One definition is that science is an attempt to answer the question 'why?' What is the reason things behave as they do? After all, the word 'science' derives from the Latin *scire*, 'to know'. However, religion could just as easily match this definition, so there must be more to it than that. An important aspect of this 'more' lies in how science works, or the basis on which it tries to provide the answer to the question 'why?' This basis is, of course, data, or evidence. Science explicitly collects data and weighs up the various proposed answers according to how well they explain the observed data. Moreover, there is a strategy to the collection of these data.

Two thousand years ago, the Greeks stressed the role of reason, thought, and deduction, exemplified by Aristotle's syllogistic reasoning (if A implies B, and B implies C, then A implies C), in understanding nature. With the rediscovery of ancient Greek writings, interest was revived in the thirteenth century by people such as Thomas Aquinas. He and others, called *schoolmen*, used these principles in an attempt to provide a solid foundation for Christian beliefs. By the early sixteenth century, reasoning, coupled with respect for the ancient sages, formed the basic model for human efforts to understand nature and the universe. Richard Jones, in his 1961 book *Ancients and Moderns*, puts it nicely: 'Dazzled by the recovered light of the past, the Elizabethans so invested the ancients with the robes of authority that the latter became oracles, to question which bordered upon sacrilege.' In medical training, for example, the key tool was a library of old books, often without even the need to have seen a patient.

But the cat was out of the bag: the powers of deductive argument were too much to be used for just the one purpose of bolstering Christian beliefs. In particular, they could be used to argue against beliefs founded solely on authority. Furthermore, there was a growing awareness of situations in which the ancients' assertions contradicted observation, so that various thinkers began to question this acquiescence, stressing the importance of observation and experiment over accepted wisdom. I shall mention just three key players: Paracelsus, William Gilbert and Francis Bacon, two of whom we have already encountered.

Paracelsus (1493–1541) studied alchemy, surgery, and medicine at the University of Basel, but was forced to leave Basel after a disagreement about his studies of necromancy. After adventures which took him to many countries, he returned to the University of Basel in 1526. Medical teaching at this time was based on the teaching of the Greek Galen (AD 129–c.210). Galen had synthesized previous medical thought and sought to construct a unified system, but Paracelsus criticized this system and the awe in which it was held, going so far as to burn Galen's works (which did not go down too well with the authorities, and he was forced, for a second time, to leave the university of Basel). He proposed an alternative approach, putting a stress on observation. Despite his emphasis on the doctrine of signatures, discussed earlier, the Canadian philosopher Manly Hall (who has written about Paracelsus in Hall, 1990) described him as 'the precursor of chemical pharmacology and therapeutics and the most original medical thinker of the sixteenth century'.

William Gilbert (1544–1603) is immortalized for his early investigations into magnetism and electrostatics, described in *De Magnete* (1600). He explicitly accused the ancients of making errors: 'it is permitted us to philosophise freely and with the same liberty which the Egyptians, Greeks, and Latins formerly used in publishing their dogmas: whereof very many errors have been handed down in turn to later authors: and in which smatterers still persist, and wander as though in perpetual darkness,' and criticised those who 'stubbornly ground their opinions on the sentiments of the ancients' for simply accepting what the old books say, rather than making observations of their own.

The final member of the trio is Francis Bacon.

Francis Bacon is often regarded as the father of 'experimental philosophy', the term used in the seventeenth century to indicate the notion of overtly collecting data to explore a hypothesis and what we might now regard as one of the cornerstones of modern science. Like Gilbert, Bacon was unhappy with the unquestioning acceptance of the teachings of ancient authorities, recognizing that these contained errors and imperfections. He saw these ancient teachings as discouraging mankind from trying anything new, and as holding up progress. In particular, he objected to the ways the Greeks had relied on reason, rather than on observation. Bacon was an astute man, and was aware of the currents of his time. He stressed (again like Gilbert) that he was not saying that the ancients were stupid – that would be risking too much in the context of the then intellectual and philosophical climate. On the contrary, he said, the ancients were extremely clever, but they simply had not adopted the strategy of collecting data to advance their knowledge.

Bacon had to overcome the attitude that human dignity suffered from involvement with lowly material things. He and his followers tackled this attitude head on by ridiculing it. Funnily enough, we find exactly the same kind of attitude still reflected in certain modern sciences. In particular, in those sciences and technologies which have a mathematical basis, such as physics, statistics, and computer science, one often finds that the most highly regarded pieces of work are amongst the most abstract, and that the most highly regarded journals are those which publish the more theoretical and abstruse material, far from direct practical applications. This is despite the fact that those articles most directly concerned with practical applications are likely to yield the most striking benefit. Plus ça change …

Bacon adopted a fundamentally cautious attitude, and again one might describe this as a characteristic of science. Science seeks an explanation of the data, but should not claim that it has the 'true' explanation, because it is the essence of science that new data may be found to contradict this explanation, and that better explanations may be found. Here again Bacon was very modern – and perhaps idealistic. Moreover, whereas he described the approach of the ancients as being to leap from a few empirical observations to the highest level of general theory, he proposed a gradual incremental approach to building up understanding. At the lowest level, the generalizations are little different from the raw data themselves, but these are gradually combined into higher level concepts, ideas and theories, over an extended period of time and with the collection of more data, until high level ideas eventually emerge.

Perhaps one of the most interesting aspects of Bacon's philosophy is that he stressed the notion of contradictory data. Data supporting theories are all very well, but the true test of a theory is the idea of refutation: of finding experimental data which tell us a theory cannot be right. Bacon's ideas are probably slightly different from modern ideas (or perhaps less precisely formulated), but the basic principle is the same. He recognized the vast amounts of effort, and therefore cost, which would be involved, and suggested that the State should be involved (it should be a 'royal work'): 'I take it that all these things are to be held possible and performable, which may be done by some persons, though not by one alone; and which may be done in the succession of ages, though not in one man's life; and lastly, which may be done by public designation and expense, though not by private means and endeavour' (Spedding *et al.*, 1879–1890).

Bacon had a great influence on the early development of what we can properly call science, but his reputation suffered something of a decline after the mid-nineteenth century. In part this was due to the continued growth of the 'mechanical philosophy', with its implicit underpinning that scientific laws were best expressed in mathematical terms. More recently, however, with a broader appreciation of what science is, and the differences between the different sciences, his reputation has begun to recover. Fundamentally, all sciences depend on data, and this was the essence of Bacon's point. He was also responsible for a reduction in the intellectual attributes regarded as necessary to carry out investigations into 'natural philosophy'. Instead of the hard-won knowledge of ancient texts, instead of the apparent clarity with which the ancients had argued their cases, what was needed was an ability to physically manipulate objects and substances: an ability to collect data. Moreover, to carry out the large numbers of experiments a large number of experimenters were needed, and this could only be achieved by relaxing standards. Of course, the truth is that careful experimentation, collecting accurate data, requires a high level of skill of its own type.

If thinkers such as Gilbert and Bacon created the roots of scientific advancement, it was the seventeenth century which saw it blossom.

Built on the principles of practical investigation and observation of phenomena, that is, on the gathering of data, this period saw the foundations laid for our modern understanding of the physical world. From our current perspective, this was a period of intellectual giants. It includes names such as Isaac Newton (1642–1727), Robert Hooke (1635–1703), Robert Boyle (1627–1691), Edmond Halley (1656–1742), John Flamsteed (1646–1719), Gottfried Leibniz (1646–1716), and Christiaan Huygens (1629–1693). The period is often termed the 'scientific revolution', and to describe it as revolutionary is no exaggeration. Of course it is true, as we have seen above, that the seeds of this revolution had been laid far back, even during the Middle Ages, with for example, the ideas of Friar Roger Bacon, but they certainly came to fruition in this period.

Science: the foundations

There are two ways of looking at historical developments. One is to look back, from our current position and, adopting our current perspective, to cast the developments into our modern framework. From this perspective, Newton laid the foundations of modern physics. Another way is to cast oneself back to the context of the time, and look at the developments of the period not as the precursors to what was to follow, but as the culmination of what had gone before. From this perspective, for example, Newton's wrestling with alchemy and religious doctrine, as well as his wrestling with gravitation, optics and differential calculus, all make perfectly good sense. As this book is concerned with the results of these developments, I will tend to favour the first of these perspectives. In doing so, I am not trying to judge the past through the eyes of the present, but merely saying that it was such-and-such a development that led to where we are now. I am interested in where and when things happened that led to our current views on data, evidence and how we should use these to understand the world – that is, on what science is and how science should work.

The word 'science' is a relatively new word, being a nineteenth-century creation. We can talk about Newton, Hooke, Boyle, etc. as being early scientists, but if we do so, we are certainly describing them in our terms. In a sense, these people, building on the principles outlined by Francis Bacon, were the creators of science and 'the scientific method', so it would be inappropriate to regard them as working within a pre-existing scientific framework. The term natural philosophy already existed. The word 'natural' here refers to 'nature', and contrasts natural philosophy with moral philosophy, which was concerned with ethics. Natural philosophy placed more emphasis on deduction and the role of authoritative sources than on observation. Moreover, there were various other disciplines, for example, astronomy and pharmacology that nowadays come under the remit of science, which previously were separate.

By 1642, the year in which Newton was born, the authority of the ancients had been broken, not least by two key advances in understanding. One was the recognition of the orbiting and rotation of the planets, and the other was the discovery of the circulation of the blood, and both were based on observation.

Political events led to London and Paris becoming the centre for these developments (even the Dutchman Christiaan Huygens spent most of his productive time in Paris), and two scientific societies were established that were to go on to assume pre-eminence amongst such bodies and which still exist today: the Royal Society of London (1662) and the Académie Royale des Sciences (1666). The first of these was stimulated by the end of the English Civil War in 1645, when a group of early supporters of the new philosophy met to discuss their experiments. In his early history of the Royal Society in 1667, Bishop Thomas Sprat wrote:

> For such a candid, and unpassionate company, as that was, and for such a gloomy season, what could have been a fitter Subject to pitch upon than *Natural Philosophy*?
>
> ... *that* never separates us into mortal Factions; that gives us room to differ, without animosity; and permits us, to raise contrary imaginations upon it, without any danger of a *Civil War*.
>
> Their meetings were as frequent, as their affairs permitted: their proceedings rather by action, then discourse; chiefly attending

some particular Trials, in *Chymistry*, or *Mechanicks*: they had no
Rules nor Method fix'd: their intention was more, to communicate
to each other, their discoveries, which they could make in so
narrow a compass, than an united, constant, or regular inquisition.

(Sprat, 1959)

It has to be said that the high-minded aims of the Society were not
met with a corresponding level of regard by others. Parts of *Gulliver's
Travels* are based on the Royal Society (with, for example, its description of men trying to 'extract light from cucumbers'). Note Sprat's
reference to 'action rather than discourse'; that is, the experimental
philosophy of getting one's hands dirty, collecting data and observing how things actually behave, rather than speculative discussion.
Of course, Sprat was presenting an idealized and optimistic view of
things. Modern science is an intensely competitive business and it
was not long before conflicts enveloped members of the Royal
Society. Examples are the battle between Newton and Liebniz over
the first invention of the calculus, the disagreements between
Newton and Hooke on the nature of light and of celestial motion,
and the argument between Newton and Flamsteed over Flamsteed's
data. (It is no coincidence that the name Newton appears in all of
these disputes. We shall return to this later.)

It is very clear from the early history of these societies that
'science' was regarded as part of the culture of society. This is
perhaps in contrast to more recent perceptions, exemplified by the
peculiar portrayal of white-coated scientists in popular films. In
England, with the Royal Society, another major difference between
then and now, and perhaps one related to the way in which science is
regarded, is that these early 'proto-scientists' were entirely self-
financing (membership then was one shilling per member per
week). At this stage, things had not reached Bacon's vision of State-
funded research. In France, on the other hand, the Academy of
Science was royally supported – a counterpart to the Academies of
Literature and Fine Art. The advantage of being self-financing was
that there were no restrictions on what might be investigated,
although, as Robert Hooke put it in a draft introduction to the

Statutes of the Royal Society written in 1663: The business of the Royal Society is: 'To improve the knowledge of naturall things, and all useful Arts, Manufactures, Mechanick practices, Engynes and Inventions by Experiment – (not medding with Divinity, Metaphysics, Morals, Politics, Grammar, Rhetorick, or Logicks).'

A central role in the early Royal Society was played by Robert Boyle, who declined an invitation to become its president in 1680. He is still remembered 300 years later for his eponymous gas laws. It is perhaps especially interesting that he was a religious man, seeing science as revealing God's work rather than being in opposition to or contradiction of it. He was also wealthy, a son of Richard Boyle, the first Earl of Cork, and he employed as an assistant one Robert Hooke. Hooke was not wealthy, and required financial support to enable him to pursue his scientific interests. Fortunately, Hooke was a masterful experimenter in contrast to Boyle. J. D. Bernal, in the book *Science in History* (1954), describes Hooke as 'the greatest experimental physicist before Faraday', and he has been the subject of several recent biographies (Inwood, 2002; Bennett *et al.*, 2003; Jardine, 2003). Like Boyle, he too discovered physical laws which are still named after him to this day. Hooke's investigations nicely illustrate the breadth of these early scientists. He made discoveries in physics (e.g. Hooke's Law, that extension is proportional to force, as well as developing ideas about gravity, even originating the inverse square law), mechanics (inventing the balance wheel), microscopics (he wrote *Micrographia*, describing discoveries made with the microscope, and which Samuel Pepys described as 'the most ingenious book that I ever read in my life'), and a host of other areas, not least in his plan being chosen to rebuild London after it was destroyed in the Great Fire of 1666.

Amongst all these early proto-scientists, however, one reigns preeminent. This is, of course, Sir Isaac Newton, whose picture used to appear on the British one pound banknote. Like Hooke and Boyle, Newton's Laws are still familiar 300 years after his death. These laws describe the way objects move under forces, and unite, in incredibly simple form, the reason why apples fall towards the ground, why the moon orbits the earth, and why galaxies take the shapes they do. It was Newton's equations that were used when men first went to the moon. Like others of his age, Newton's interests were much wider than the narrow area of kinematics. He also invented the area of mathematics known as the calculus, although he did not publish it until forced to by the fact that the German mathematician Leibniz had invented something very similar. This represents an advance just as fundamental as the discovery of gravity. The calculus underlies modern engineering, electronics, physics, even the theory of option trading in the financial markets. This single invention alone has revolutionized human history. But Newton made even more than those two major breakthroughs. He also made major advances in optics, and was responsible for the discovery that white light was a mixture of light of all colours, which could be separated using a prism. This observation led him to recognize that *reflecting* telescopes (in which light is reflected from a curved mirror) would produce higher quality images than classical *refracting* telescopes (which pass light through a lens), since the latter will tend to separate the light according to its constituent wavelengths. Although the principle of such telescopes had been known for a long time, the devices had proved very difficult to make. Newton, however, apparently working alone, constructed one in 1669, devising the metal alloy for the mirror and casting and grinding it himself. It was presented to the Royal Society, an organization with which Newton had previously had no involvement, in 1671, to wide acclaim. He was elected to the Society in 1672. Later in Newton's life, John Conduitt asked him where he had had his telescope made: 'he said he made it himself, and when I asked him where he got his tools he said he made them himself and laughing added that if I had stayed for other people to make my tools and things for me, I would have never made anything of it.' This is from Conduitt's memorandum of 31 August, 1726, in John Maynard Keynes manuscript 130, 10 fol. 3–3v.

Keynes, a great economist, purchased a collection of Newton's papers at an auction in 1936 and left it to King's College Cambridge when he died. It was these papers that revealed that Newton's investigations had been much wider than had been described by earlier biographers. These earlier biographers (in particular, William Stukeley, who knew Newton and wrote *Memoirs of Sir Isaac Newton's Life*, soon after his death, and Sir David Brewster, who wrote *Memoirs of the Life, Writings, and Discoveries of Sir Isaac Newton*) had painted a grossly over-idealized image of Newton as someone whose thoughts were guided by pure reason, and who had an unerring ability to steer toward the rational and avoid the irrational. The fact is, however, that Newton regarded his investigations into alchemy and theology as equally important as his scientific investigations. Perhaps unsurprisingly, his non-scientific work was just as original and deep, within its own context, as his scientific work. Referring to the rediscovery of these aspects of Newton's work, Keynes said:

> Newton was not the first of the age of reason. He was the last of the magicians, the last of the Babylonians and Sumerians, the last great mind which looked out on the visible and intellectual world with the same eyes as those who began to build our intellectual inheritance rather less than 10,000 years ago. Isaac Newton, a posthumous child born with no father on Christmas Day, 1642, was the last wonder-child to whom the Magi could do sincere and appropriate homage.
>
> (Keynes, 1947)

Some insight into Newton's personality is perhaps gauged by the reaction to the lectures he was obliged to give when he was at Cambridge University. For many of them he had no audience at all, but, duty bound, he nevertheless delivered them – to the empty lecture theatre (although, in fairness, it has to be said, he often halved the length of the lecture from the nominal 30 minutes). He was intensely secretive about and also proud of his work, which led him into bitter conflicts with others of the age.

With Leibniz, it was an extended priority dispute about who had invented the calculus first and, from Newton's side, whether Leibniz had stolen the ideas from him. This seems not to be the case. Leibniz was a brilliant man in his own right (sometimes even described as 'the Continental Newton') and it seems to be a genuine instance of parallel development. In fact, such things are quite common in the history of science and in modern science. When a theory's time has come, it is recognized by more than one person. This is as true of gravity, relativity, evolution and the structure of DNA as it is of the calculus. (It would be a mistake to believe that science is a gentle and gentlemanly pursuit, and the feuds, disputes, and fights that occur within it are pursued with as much vehemence and animosity as any in the world of business.)

Newton's disputes with Hooke probably owe at least as much to differences in their personalities as in their science. Although both had suffered childhood hardships, and had had to fund their university studies by acting as servants to other richer students, they developed in very different ways, with Newton becoming more introverted and hypersensitive to criticism, and Hooke becoming more extroverted and outspoken. An early dispute hinged around conflicting theories of light. Modern science uses a process known as *peer review*, in which papers submitted to journals are examined by referees, other scientific experts in the same area and who recommend publication or rejection, to see if the arguments are consistent and valid. A similar process operated in the early days of the Royal Society, and Hooke was tasked with examining and commenting on Newton's *Theory of Light and Colours*, his first submission to the Society. Unfortunately, the ideas in this work disagreed with Hooke's own conclusions about the nature of light. He wrote: 'as to his hypothesis of solving the phenomenon of colours thereby I confess I cannot see any undeniable argument to convince me of the certainty thereof'. (from Turnbull 1959, pp. 110–111). This did not go down well with Newton, and an escalating angry correspondence ensued. Eventually, in a long and detailed letter examining every one of Hooke's objections to his theory, Newton won the argument. Of course, he did not make a friend, and the antagonism between the two lasted until Hooke died, resurfacing from time to time in other contexts.

There are also similarities between the early lives of John Flamsteed and Newton, with Flamsteed's mother dying when he was three. Flamsteed was appointed Astronomical Observer by King

Charles II, a post which became that of Astronomer Royal. From the Greenwich Observatory, Flamsteed made accurate measurements of the positions of the stars and planets, and provided Newton with data, especially relating to the moon. Unfortunately, the perfectionist Newton demanded perfect data, and when mistakes were found he was critical. The dispute escalated, doubtless driven in large part by Newton's insensitive approach to handling others, and ended only when Newton's interests moved elsewhere, although later the animosity between the two resurfaced when they clashed over the publication of a catalogue of heavenly objects.

Yet another correspondent with whom Newton disputed was the French Jesuit Ignance Gaston Pardies. In his reply to a query from Pardies about the cause of the splitting of light into colours by a prism, Newton replied: 'For the best and safest method of philosophising seems to be, first to inquire diligently into the properties of things, and establishing those properties by experiments and then to proceed more slowly to hypotheses for explanations of them.' He could not have put the Baconian philosophy and the case for collecting data more transparently.

Insensitive, irascible, quick to take offence, unable to understand others, Newton was clearly an unprepossessing character, but genius, perhaps the greatest humanity has ever seen, he also doubtless was. One of the key things that distinguishes him from many of the other researchers of the time was that he combined consummate experimental skills with superb mathematical ability. It also has to be said that the fact that he outlived Hooke enabled him to cement his reputation at the expense of anything that Hooke had achieved.

What is science?

We have seen that, around the period of the sixteenth century, there was an increased awareness of the key role that observation, experiment and above all, data, must play in understanding nature. However, there is more to science than the simple collection of facts. (Railway timetables and stellar catalogues contain collections of facts, but in themselves, they are hardly science.) Science also attempts to explain those facts, to create some unifying 'theory' that summarizes them and ties them together, so that one can understand why things happen the way they do. The aim of science is to understand.

The ancient Greek approach to understanding nature can be characterized as being one of *deduction*. In a deductive process, one begins with certain basic truths or assumptions (axioms) and deduces the consequences of them. The Greek mathematicians and geometers, such as Pythagoras and Euclid, epitomize this, regarding their axioms as self-evident. (Later mathematicians disagreed, and major advances resulted from changing the axioms.) But science is different. In science, one is not seeking to deduce the consequences of known properties, but rather to deduce the properties from observed facts. What is needed is really the opposite of deduction: *induction*. In induction, we observe some facts or instances of a phenomenon, and try to derive the nature of the phenomenon from those observations. Francis Bacon's great contribution was that he recognized the need for and promoted the use of induction as opposed to deduction.

Science is often said to be defined by 'the scientific method', so we must ask what this means. It obviously describes the process by which the mass of facts are reduced to a simple theory, the process by which the mechanisms underlying the generation of the facts are teased out. But how does it work: what is this 'method'?

In a sense, the scientific method does not exist at all, but rather is an abstract idealization. It is a description of how we *would* proceed in our voyage of discovery in an imaginary world uncorrupted by distorted data, scientific competition, financial concerns, time limitations, ethical considerations, and so on. In such a context, the scientific method consists of the following steps:

- collecting data about the phenomenon being studied;
- conjecturing some explanation ('theory', 'hypothesis', etc.) for the data;
- using this proposed explanation to predict what future data will look like;

- collecting more data;
- seeing if it fits the predictions; and
- oscillating round and round these steps, gradually refining one's explanation.

If the data do fit the predictions, then one might regard this as evidence supporting the proposed explanation. If the data do not fit the predictions then (in our imaginary ideal world) it is an indication that there is something wrong with the proposed explanation. In this case, one will go back and modify the proposed explanation (adjust the theory) or come up with some alternative new explanation which does fit the new data, and which also fits the data one has already collected, of course. Then one will make some predictions based on this new explanation, collect some more data, see if it fits the predictions, and repeat the cycle. By repeating this cycle again and again, one hopes to get nearer and nearer to the 'true' explanation for the phenomenon under investigation. Certainly one is constructing theories which provide better explanations, in that they correctly explain more data and make more accurate predictions.

There are, however, several aspects of this idealization which merit closer scrutiny.

First, although the list of steps above started with data, things are not really that straightforward. Admittedly, this was what Bacon proposed; that we should first collect a large number of facts, unencumbered by the way they might fit together, and only then try to fit them into a theory, and many early scientists saw this as a description of what they were trying to do. Charles Darwin said of himself that 'without any theory [he] collected facts on a wholesale scale'. In his autobiography, he wrote 'After I returned to England it appeared to me that ... by collecting all facts which bore in any way on the variation of animals and plants under domestication and nature, some light might perhaps be thrown on the whole subject. My first notebook was opened in July 1837. I worked on true Baconian principles, and without any theory collected facts on a wholesale scale.'

In fact, of course, research does not begin in a vacuum – all researchers start from a context of some ideas, previous theories, general background knowledge, and so on. The leapfrogging act of data collection, followed by theory construction, followed by data collection, followed by theory construction, and so on, has no obvious beginning. Apart from anything else, it would be extraordinarily inefficient to discard all that had gone before: most scientific advances build on earlier work, rather than starting from scratch. Furthermore, researchers are as impatient as anyone else, and are keen to formulate their ideas as soon as sufficient data becomes available on which to do so. Since, as we discuss below, there is a premium on being the first to discover something, there is also implicit peer group pressure not to hang around for too long before making a conjecture. Of course, sometimes this can spectacularly backfire, and scientists who rush in too soon can end up looking foolish when new data refutes their proposed explanation. This is one reason why the Nobel Prizes in science are generally awarded many years after the breakthrough they are recognizing: only time can tell if they really are advances rather than false turnings.

In general, then, it is all very well to talk about collecting data in the abstract, but no data are completely divorced from some theory. The work of Nobel Prize-winning physicist Robert Millikan, discussed further in Chapter 7, illustrates this nicely. He supported the theory that electricity was fundamentally particulate in nature, and that therefore there was a smallest possible charge. He carried out a series of highly sensitive and sophisticated experiments to verify this, seeking to demonstrate that all the tiny charges he could measure were multiples of some smallest value. However, his notebooks show that he was selective about which results he regarded as accurate, and which could therefore be published and analysed. He could only do this in the context of his theory: how he interpreted and regarded his data are a function of his beliefs.

Here is another, more elaborate example. This is how the geophysicist Sir Alan Cook described how data describing the intensity of light emitted by a black body are collected:

The light falls on semiconductor detector that generates an electrical signal that causes another semiconductor device, a digital voltmeter, to emit a train of electrical pulses that set up a certain state in the memory circuits (more semiconductors) of a digital computer. The computer also receives a train of signals from a second complex of electronic devices that purports to measure the temperature of the black body. Finally, the computer, having itself issued electronic instructions for changing the temperature, calculates a relation between temperature and light intensity.

(Cook, 1994).

Sir Alan went on to remark: 'It would be difficult to argue that the result is independent of theory.' Although it may not look like it, even so simple a thing as measuring the length of something involves an implicit theoretical component: as we saw in Chapter 2, one has to choose what is to be the unit length (inch, meter, pencil, etc.) and then define some way of comparing the object to be measured with the unit length (e.g. 'starting from one end of the object to be measured, lay copies of the unit length, end to end, until they reach the final end of the object to be measured. The length of the object is then the number of copies of the unit length'). Clearly there is more to data than meets the eye.

Second, data are seldom perfect: they have errors, distortions, missing values, and so on. (These sorts of problems are the subject of Chapter 6.) So, when we propose a theory and then collect data to test it, and find that the data do not match the predictions, is it not possible that this is simply because our data were poor? If we had measured things more accurately, if crucial pieces of data had not been missing, might we not have found that they did match the theory? More generally, how do we decide if a match between data and prediction is close enough to support our theory? If, based on my theory, I predict that the pressure of a fixed volume of gas will increase by three psi (pounds per square inch) when I increase the temperature by a certain amount, but in fact I find the increase is 3.1 psi, is this close enough? What about 3.01 psi? We are unlikely to observe *exactly* 3.0 psi, so does a result of 3.1 support or refute my theory? What all of this means is that the business of deciding whether data match the predictions or not is often not as simple as one might wish and, in fact, often when a mismatch does occur, scientists suspect the data rather than the theory. As Nobel Prize winner Paul Dirac (1902–1984), one of the founders of quantum physics, put it: 'It is more important to have beauty in one's equations than to have them fit [the] experiment … The discrepancy may well be due to minor features … that will get cleared up with further developments.' Science is not merely a question of turning an experimental handle: it requires careful thought and creative investigation.

In this vein, new sources of data and observations arising from new types of instruments may not be accepted immediately. While, in retrospect, they might open up new ways of looking at things, revealing entirely unsuspected worlds about us, they could also conceivably distort things, leading to fundamental misconceptions about nature – think of the distorting mirrors in a fairground. Not all of the investigators in Galileo's time accepted the telescope as leading to legitimate observations, and some refused to use it. But the telescope had the advantage of clear practical value: it was very useful in determining planetary and stellar positions, and hence in navigation. The optical instrument at the other end of the scale, the microscope, did not have any such obvious practical merits. It took a leap of faith to acknowledge that the tiny wriggling creatures Antoni van Leeuwenhoek (1632–1723) saw in drops of water were real, or that the capillaries connecting arteries and veins, seen by Marcello Malpighi (1628–1694), represented real physical structures. After the beginnings of the microscope, and the early discoveries by such as Malpighi, Swammerdam, Hooke, Grew and Leeuwenhoek, the eighteenth century saw relatively little serious scientific use of this device, and it was not until the nineteenth century that things picked up again. Although there are always technical difficulties in learning to use a new instrument to best effect, in the case of the microscope perhaps a greater problem was a philosophical approach to medicine which centred on symptoms rather than biology. It meant, in particular, that the anatomical discoveries facilitated by the microscope did not translate into immediate medical advances. It was all a question of what data were regarded as relevant. For example, with symptom patterns as central to the basis of their understanding of medicine, Thomas Sydenham (1624–89) and John Locke (1632–1704) maintained that a relatively superficial understanding of anatomy was sufficient. When Hans Sloane, introduced as a 'ripe scholar, a good botanist, and skillful anatomist,' arrived to study with Sydenham, the latter responded: 'This is all very well, but it won't do – Anatomy – Botany, Nonsense! Sir, I know an old woman in Covent Garden who understands Botany better, and as for Anatomy, my butcher can dissect a joint full as well; no, young man, all this is stuff: you must go to the bedside, it is there alone you can learn disease' (Payne, 1900). Sydenham and Locke were rejecting the notion that careful exploration with instruments can shed much light on disease, and were arguing instead that careful observation of diseases in their natural state is sufficient – they are deciding what data should be used on the basis of their prior theories.

It is possible that people would nowadays be less suspicious of new sources of data, because of generations of experience with instruments that dramatically extend man's senses. We now have CAT, MRI, PET and SPECT scans as well as radio telescopes, infra-red imagers, sonar systems and synthetic aperture radar, not to mention biological probes arising from work in genomics and proteomics.

Third, the world is immensely complex. Even apart from errors in the data, it would be asking a bit much to expect a theory to explain *everything* perfectly. Take the case of a falling object again. We have our theory of gravitation to explain the object's motion in terms of the gravitational attraction between the object and the earth, and it does an excellent job. Indeed, the mathematical form of our theory even explains the rate at which the object gains speed as it falls towards the Earth. But closer examination, more accurate measurement, shows that our theory is not a perfect match for the data. It may not be a perfect match because, perhaps, the force between two bodies may not be exactly an inverse square law (after all, the bodies are not point masses), so that our theory needs refinement. It is certainly not a perfect match, because the theory has failed to take account of air resistance as the body falls, perhaps even wind pushing the body to the side, the effect of the attraction of other bodies, or even the pressure of light falling on the body. Once again, we have to decide whether or not our theory should be rejected because the match between data and theory is not perfect. I repeat: the world is a complex place, and no theory can explain everything.

In fact, this was nicely illustrated recently by an article in the UK's *Independent* newspaper (8 December, 2003). The article reported Allen Roses, from Duke University in North Carolina and the worldwide vice-president of genetics at the pharmaceutical giant GlaxoSmithKline, as saying 'fewer than half of the patients prescribed some of the most expensive drugs actually derived any benefit from them,' and the headline said 'Glaxo chief: our drugs do not work on most patients.' The fact is that different people react differently to the same drug. The effect will depend on obvious things such as body weight and sex, but also on genetic make-up, previous exposure to disease and treatments, other concurrent treatments and a whole host of different factors. To expect the wide variety of human beings to respond in an identical way to the same dose of a drug is simply naive, and Dr Roses was merely making that point. In fact, science is making substantial advances in this area – the theories are developing dramatically and are now able to take into account more and more of the factors which differentiate between people. In particular, recent developments mean that in the near future we will be able to identify much more accurately who is likely to benefit from a particular drug on the basis of their genetic make-up. This will save a great deal of money, as well as stopping the unnecessary exposure of 'non-responders' to drugs from which they will not benefit.

Fourth, data conforming to the predictions are often easy to come by. Every time I release a ball and observe that it falls downwards, it supports a theory which says that all objects in the universe attract each other, and I can easily collect a vast amount of data like this. On the other hand, this is a very weak sort of support. Much better would be a tougher test: does the motion of the moon about the Earth lead to objects on the Earth trying to follow the moon because of its attraction? (It does, of course, as is evident in the tides.) In general, the tougher the better. The more searching the challenge our theory overcomes, the more confidence we can have in it. Tests that are easy to pass – that most theories would pass – are fairly useless. Likewise, ambiguous tests, those which do not give a clear outcome ('*perhaps* it supports the theory'), are of limited value.

The core of science, the root of 'what science is', lies in this notion of *falsifiability*. That data collection procedures, that is experiments, can be devised such that, if the data do not match the predictions of the theory, then the theory is wanting (subject, of course, to the difficulties associated with the imperfect world in which we live). In contrast, if a theory could explain all possible outcomes, so that no possible outcome could lead to its rejection, then it is not science. If one can retrospectively justify any outcome as being in accordance with a theory, then that theory does not have much intellectual meat. It is useless for prediction, since according to it anything can happen. Psychoanalysis, discussed briefly in Chapter 7, has been accused of being non-science in this sense: whatever data were collected about human behaviour, psychoanalysis could explain it. This situation (whether it is true of psychoanalysis or not) is the very antithesis of science.

This is an absolutely key point. Science and scientific theories are, above all, *contingent*, in the sense that they are determined by the data they seek to explain. At any time, new data could come along that contradicts the theories, requiring them to be modified. *Science cannot provide certainty*. It merely provides the best explanation so far, for the data which have been collected so far. Only religion provides certainty, and that certainty is not based on data, but on faith. (And, one is tempted to ask, if religion offers certainty, how can it be that there are conflicting religions which hold different things to be true?) We can put this another way. Science does not claim to be infallible, rather it accepts and encompasses fallibility as part of its driving force, explicitly pitting its theories against new data in an attempt to test their veracity. Other forms of knowledge, such as religion, politics, literary theory, aesthetics, magic, etc., typically lack this intrinsic testability, or, at least, are not tested in this way, and so necessarily have a major subjective component. In a speech given in 1918 (*Science as a Vocation*), the great sociologist Max Weber said:

> science has a fate that profoundly distinguishes it from artistic work ....A work of art which is genuine 'fulfilment' is never surpassed; it will never be antiquated. Individuals may differ in appreciating the personal significance of works of art, but no one will ever be able to say of such a work that it is 'outstripped' by another work which is also 'fulfillment' ... In science, each of us knows that what he has accomplished will be antiquated in ten, twenty, fifty years. That is the fate to which science is subjected; it is the very *meaning* of scientific work, to which it is devoted in a quite specific sense, as compared with other spheres of culture for which in general the same holds. Every scientific 'fulfilment' raises new 'questions'; it *asks* to be 'surpassed' and outdated. Whoever wishes to serve science has to resign himself to this fact ... We cannot work without hoping that others will advance further than we have.

This business of fallibility is important because it is something that is often, perhaps quite naturally, misunderstood by the non-scientific community. For example, one day one reads that coffee is bad for you, and the very next day that it is good. This sort of thing leads to confusion and doubt about the value of science. Surely science is supposed to provide *the* answer, based on hard data, so doesn't this constant changing of minds means that scientists don't really know what they are talking about? Why, then, should we trust them? But we have seen that it is precisely *because* scientists are prepared to change their minds that we should trust them. The theories represent the best interpretation of the evidence (data) which have been collected so far. When new data comes along, it is perfectly natural, indeed only reasonable, that the theories may change. What would be the alternative? That one stuck to one's old theories in the light of contradictory data? I am reminded of John Maynard Keynes' response to a critic accusing him of being inconsistent: 'When somebody persuades me that I am wrong,' he said, 'I change my mind. What do you do?'

Even those who should know better fall prey to this misconception that science must get it right first time (or, at least, that science must get it right 'this time'), not appreciating that 'this time' is merely the latest in a series of more refined theories, each better than those that have preceded it (in the sense that it explains more data, or explains the data better), but that one should not expect it to be the last theory (the 'last theory' will only come when scientific progress stops, and that will only happen when humanity itself stops). A nice (or, perhaps, profoundly depressing, depending on how you look at it) example of this was given by Michael Meacher, the UK's Minister for the Environment from 1997 to 2003, in an article in *The Times* (29 April, 2004), who wrote:

> Science quite often gets things wrong. Biologists initially refused to accept that power stations could kill fish or trees hundreds of miles away in Scandinavia; later the idea was universally accepted. Scientists did not originally agree that cholorfluorocarbons (CFCs) were destroying the ozone layer; but when the industry – ICI and DuPont – abruptly changed sides in 1987, ministers and scientists soon lined up with them. The Lawther working party of Government scientists roundly rejected that health-damaging levels of lead in the blood came mainly from vehicle exhausts, only to find that blood-lead levels fell 70 per cent after lead-free petrol was introduced. The Southwood committee of BSE scientists insisted in 1989 that scrapie in cattle could not cross the species barrier, only to find by 1996 that it did just that.

Well, yes. Adopting a working hypothesis, while continuing to investigate it, and test it against new evidence is what science is all about. The essence of science is the acknowledgement that one's knowledge is contingent and might well change as new information

UNIVERSITY OF HERTFORDSHIRE LRC

becomes available. The core of science is that the current best work-ing hypothesis (in the cases cited by Meacher, that power stations could not kill hundreds of miles away, that CFCs were safe, that blood lead did not come from petrol, and that scrapie did not cross the species barrier) are tested and beaten with every experimental stick one can think of, in order to test their robustness. And when they fail, they are replaced by something more robust. As Keynes might have put it, what is the alternative? To stick with one's initial beliefs in the light of evidence that one is wrong? All of this means, of course, that no scientist can ever be certain. There can be no guarantee that new evidence will not come up which refutes one's current theory. Certainty, as I noted above, is beyond the realm of science.

Once again, we should really qualify the above with the reminder that we do not live in a perfect world. The truth is that human beings make science, and human beings have their foibles and weaknesses. Moreover, scientific research is a competitive arena, with reputa-tions hinging on the success of one's theories. Whoever invents the dominant theory will find kudos, peer regard, prizes and even finan-cial rewards. Given this, it is hardly surprising to find that there is often intense competition. Although such contests usually take place in a civilized manner through the pages of scientific journals and at conferences, the tension generated by such debates inevitably breeds its own problems. Dishonest researchers have been known to invent spurious data to support their pet theories; some examples of this are given in Chapter 7. However, this leads us on to one of the key strengths of science, and the reason that the scientific approach has revolutionized the human condition. This is that the very nature of science means that it is *self-correcting*. If incorrect theories are postulated, either because they seem to provide the best explanation for the data so far, or for other less moral reasons, then eventually the fact that they are wrong will be discovered. Data generated to test the theories will show them to be wanting; the weight of evidence will eventually wear them down. The Church imprisoned Galileo because of his refusal to accept that the Earth was the centre of the universe, but improved instruments and more data generated so much evidence that the Church had to abandon its belief. Reliance on ancient authority, as we have seen, cannot persist for long in the face of overwhelming evidence to the contrary. At a less exalted level, the so-called Piltdown Man was shown to be a fake by chemical tests some 40 years after its discovery – though there had always been doubts because the remains did not fit in with existing theories. A more recent case is given by Jan Hendrik Schön, a German physicist who worked at Bell Laboratories and who claimed to have made a major breakthrough in organic electronics which would have revolutionized future computing. Doubts were raised about this work when it was noticed that the same data appeared twice in different situations. Further investigation uncovered more duplicate data. When asked for his raw data, Schön was unable to provide it, and closer examination revealed that some of the reported data had been generated from mathematical functions in a computer. Eventually, Schön was fired, the journal *Science* withdrew eight papers and the journal *Nature* withdrew seven papers which included Schön as an author. Later replication of Schön's experiments by other researchers have failed to yield results similar to those of Schön. A key aspect of science is the careful recording of the evidence and the conditions under which the data are collected, so that the data can be replicated. A phenomenon which can be reproduced only by its original discoverer is probably not a real phenomenon at all.

The self-correcting nature of scientific progress is related to another characteristic of science: that it represents a gradual advance towards a consensus. Hypotheses are progressively conjectured and refuted as explanations of a growing body of data, and, in general, the bulk of scientists accept the dominant theories. This does not mean, of course, that other theories may not also explain the data, but it does mean that, as time progresses, so there are more data to explain: a new theory will have a harder job of explaining observed facts.

We have seen that science is a journey of exploration: an exploration of nature. Almost by definition, any such exploratory journey takes wrong turnings and has to retrace its steps to try alternative routes. At the start of their research studies, I tell my postgraduate students that research consists of two steps forward and one back, repeated time and again, slowly inching towards understanding. I also tell external bodies who wish to sponsor research that 'research is risk'; if we could guarantee the result that they would ideally like, then it would be mere development, not research.

To recapitulate, science represents an understanding of nature, achieved by a systematic process of observing how nature behaves, and then trying to summarize and explain that behaviour in terms of simplifying theories. As time progresses more data are collected, typically in a strategic manner, to test a proposed theory, so that more and more refined and powerful theories grow. Key concepts in this process are ideas of falsifiability and refutation, the notion of induction or generalization from observed data, the role of the carefully designed experiment, and – something we have not yet considered – the role of mathematics. The subsequent sections of this chapter explore these concepts in more depth.

Beating theories until they squeal

As we have seen, the Aristotelian tradition based its natural philosophy on what were regarded as self-evident truths that were supposed to be clearly apparent from everyday experience. However, not all truths are self-evident: indeed, many appear to be contrary to experience, with some being apparently paradoxical or counterintuitive without a great deal of deep thought. Likewise, many apparently self-evident truths are false. In general, it is often the case that a searching examination of a phenomenon reveals unsuspected aspects: it is easy to come up with theories to explain things at a superficial level. Recall Francis Bacon's warning: 'God forbid that we should give out a dream of our own imagination for a pattern of the world.' There are untold cases of such 'dreams of our own imagination'. For example, over the years people have suggested that lightning is a consequence of angering the gods, or the result of

Thor's hammer striking the anvil, that lightning will not strike church steeples when the bells are rung (an unfortunate theory which resulted in many deaths in mediaeval Europe), that it never strikes in the same place twice, and so on. It was not until Benjamin Franklin carried out a series of experiments in the mid-eighteenth century that lightning was recognized to be a manifestation of electricity. Franklin's famous kite experiment, in which electricity from a storm cloud ran down the damp string of a kite he was flying and made sparks fly from the bottom, again illustrates the intertwined nature of theory and data: one would probably not have thought of this experiment had one believed in the 'Thor's hammer' theory of lightning.

Careful observation and experimentation with how things behave under different conditions has led to a revolution in understanding in just about all areas. In human and animal physiology, for example, it was not until people such as Andreas Vesalius (1514–1564), at Padua, investigated things for themselves that errors were discovered in the ancient authorities. (A classic example is Vesalius's rejection of the idea that the wall between the left and right ventricles of the heart was porous – after he had studied many cadavers himself.) William Harvey (1578–1657) later studied from Vesalius's successors in Padua, acquired the tradition of carrying out experiments on animals, and went on to discover that blood circulated round the body. Again, rather than relying on tradition, he actually collected data, and others followed in his footsteps.

To make observations, and to collect data, one can either passively observe a phenomenon, or one can actively set up certain conditions under which to observe it. For the first, for example, I can sit here in my armchair, watching the clouds float by and record aspects of their shape, speed, brightness, etc. For the second, I can actively mix ingredients, heat them to a specified temperature for a given period of time, and record the results. Both approaches permit theories to be falsified even though active intervention is involved only in the second kind. In the first kind, for example, I can conjecture that brighter clouds move more swiftly than darker ones – and refute this hypothesis by observing the future behaviour of clouds. However, the second kind of data collection strategy is more powerful, because it allows one to investigate causality. For example, records show that ice cream sales and number of cases of sunburn are positively correlated – when one is large the other tends to be large, and when one is small the other tends to be small. Now it could be that eating ice cream causes sunburn or it could be that changing temperature causes both: in hot weather people go to the beaches, where they eat ice cream and are more likely to be burnt by the sun. By actively intervening – for example, by deliberately reducing the sales of ice cream on some hot days – we can test these hypotheses.

This second kind of observation, in which the conditions are actively manipulated, is an experiment. Experiments provide particularly incisive tests of hypotheses and are, in a large part, responsible for the extraordinary advances in science and more generally in human progress.

To the lay person, an experiment is often taken just to mean a description of what will happen if you do certain things in certain ways, but this is a distortion: it omits the core of the notion of 'experiment', which is that one does not know for sure what the outcome will be. The fantasy author Terry Pratchett's wonderful definition of magic is 'messing about with what you don't understand', and this certainly has the ring of an experiment to it. More formally, but in a similar vein, one could describe experiments as 'controlled procedures used to establish the facts'. At the core is always the idea that the motive for experimenting is to replace the notion of authority by a notion of evidence, which one has, or at least could have, witnessed oneself. Experiments are data collection procedures.

There is also a more technical definition of an experiment, which goes something like: an experiment is *a deliberate manipulation of things in such a way as to efficiently and accurately collect data*. So an experiment is an active intervention, in which one consciously controls some aspect of the investigation to see what effect it has on other aspects, and an experiment is also a carefully designed investigation, so that data are collected accurately, and with minimal risk of misleading results. This more formal definition will be explored in more detail in Chapter 5.

Mathematization

Perhaps one of the most striking things about modern science is that it is heavily based on mathematics. Although the popular image of mathematics is that it is about numbers – essentially advanced arithmetic – in fact it is more general than that. One might define mathematics as the study of structures involving specified sets of symbols that are combined using specified operations and that are related to each other in specified ways. Numbers, of course, are the most important example, but they are certainly not the only example. In the case of numbers, the combination operations are such things as adding and multiplying, and there are various relationships, such as relative size.

Pure mathematics is concerned only with the structures and the relationships in their own right, and is not concerned with any possible meaning. Pure mathematicians find beauty in such patterns. Applied mathematicians, on the other hand, see the structures and relationships as representing aspects of the real world: the mathematics is thus regarded as providing a *model* or an idealized description of some aspect of the world. The model is *idealized* because it will ignore many of the complications of the real world. If we build a mathematical model of the path of a shell fired from a cannon, we might take into account only its initial velocity, direction, angle from the horizontal and the pull of gravity. Of course, a more sophisticated model could try to include things such as the strength of the wind and the fact that the Earth is rotating, but no model can be sophisticated enough to include every possible influence. If we build a mathematical model of how people in a group interact, we might take into account the number of times and the length of time each pair talk to each other, but we might ignore body language and perhaps the topic of the conversations. Again, more elaborate models could be built, but one has to stop the elaboration somewhere.

In summary, mathematics is the study of abstract structures. Abstract here means the structures are just symbolic, with no reality beyond marks on a page or memory states in a computer. However, these abstract structures are often representations or models of aspects of nature. They are mathematical 'pictures' of parts of nature.

The power of mathematics comes from the fact that, if the models are sufficiently accurate, they can be used to predict how the real world will behave. If my model of the flight of a cannon shell is a good one, I can tell where it will land without going to the trouble of actually firing a shell. Conversely, I can tell with what velocity, in what direction, and at what angle I should fire the shell to *make* it land in a particular place. In 1969, a similar but much more elaborate mathematical model told NASA when to launch the Apollo 11 rocket, how much fuel it should carry, at what velocity it should travel, and in what direction it should set off from the Earth in order to arrive at the Moon's orbit, over a quarter of a million miles away, at the very time that the Moon also reached that position, thus enabling a human being to walk on an object other than the Earth for the first time ever. The model was based on data about how rockets work and on how objects behave under acceleration and gravitational pull.

The mathematical model represents an understanding of how things work, and this understanding can then be used in other situations where data collection is infeasible. We could not, for example, have launched hundreds of Apollo rockets towards the Moon just to collect information on how best to get there. We cannot experiment with the Earth's crust to see if convection currents in the Earth's mantle lead to continental drift. But using mathematical models, based on data which have been collected in simpler situations, we can produce predictions about all of these things.

We have seen that one of the things which distinguishes mankind from animals is the ability to construct complicated mental models, and manipulate them in our heads: that is, intelligence. This is taken further by the ability to build more elaborate models outside our heads, which we can then write down and manipulate manually: that is, by our ability to do mathematics. Indeed, the eminent science fiction writer Robert Heinlein once wrote: 'Anyone who cannot cope with mathematics is not fully human. At best, he is a tolerable subhuman who has learned to wear shoes, bathe, and not make messes in the house' (Heinlein, 1973). Whether or not you are prepared to go that far, it is doubtless the case that mathematical models have led to great understanding, and to a huge ability to manipulate the world about us.

As we saw in Chapter 2, the earliest uses of mathematics were motivated by trade, government, administration, etc. It was only later that mathematics also came to be regarded as central to the scientific perspective. Superficially, it is perhaps surprising that a single system should function in these two entirely distinct roles. In the one case (trade, government, etc.), mathematics is used as a system of record keeping, while in the other, it is used as a description of some aspect of the universe. Accounting is very different from description (this is nicely illustrated by the obsessive accuracy with which accountants will calculate figures – think of your taxman – as contrasted with the approximation used in models in, for example, psychology). However, things are perhaps less surprising when we look at a change which occurred during the scientific revolution in the way mathematical models were regarded.

Prior to the scientific revolution, mathematical models had been regarded as purely descriptive: they were merely convenient summaries of regularities observed in nature. Such summaries were seen as useful for prediction, but as lending no insight into what was really going on – that is, into the underlying reality. The seventeenth century saw the growth of the alternative perspective that the mathematical models did in fact reflect some underlying reality: that they were models of what *was*. Admittedly they were simplified models (idealized models, as we saw above), but they captured the core of what was really there.

A perfect example of this development is given by astronomy. All ancient civilizations had models for astronomical motions and events: such things played a more prominent role in life in those days, when the stars were not so obscured from view by artificial light. In particular, the Greeks developed mathematical models to describe heavenly motions, and by around the fourth century BC, they had developed some highly sophisticated models which put the Earth at the centre of the universe, with all the stars and planets circling around it. No flat-earthers, these. Indeed, the sophistication of their models is illustrated by the fact that Eratosthenes (276–194 BC) not only recognized that the Earth was a sphere, but he found an estimate of its diameter which was only fifteen per cent higher than the modern figure. He also calculated the distance between the Earth and the Sun, arriving at a figure only about one per cent different from the modern value. In deriving these figures, he used mathematical tools originally created for such practical needs as land surveying, but then developed further by people such as Pythagoras (c. 569–475 BC) and Euclid (c. 325–265 BC).

An example of such a model was that of Aristotle (c. 384–322 BC). He believed the sphere to be a perfect shape, and, based on this, he described a system in which the heavenly bodies were embedded in concentric spheres. It was the rotation of these which led to the (circular, of course, since this was 'perfect') motion of objects about the earth. Others (e.g. Eudoxus, c. 408–355 BC and Callippus, c. 370–310 BC) developed similar models, and the complexity of the models is illustrated by the fact that although each sphere in which a planet or star was embedded was rotating, its axis of rotation was not fixed, but was determined by opposite points in another sphere, which was itself rotating.

Later, another Greek, Ptolemy (c. AD 100–170), improved the match of the model to observations, that is to the data, by developing a mathematical model in which the planets followed motions in which circles were superimposed on circles (called epicycles – think of wheels within wheels). This was better because it modelled the times when the planets appear to move backwards compared with their usual motion and compared with the uniformly rotating stars. The model retained the idea that Man was the centre of the universe, and also that the circle was the basic form of motion. However, it departed from the perfection of the simple perfect circular model, and so, given the perspective of the time, could only be interpreted as a *description*, and not as a reflection of reality, despite the accuracy of its predictions.

Ptolemy's model held for a thousand years, surely making it one of Mankind's all-time most successful theories. It was not shaken until things began to accelerate towards the scientific revolution by the growing awareness of the importance of the need to match theory with data. In particular, in 1543 Nicolas Copernicus (1473–1543) published a book, *Oe Revolutionibus Orbius Coelestium* (Copernicus 1995), which, although similar to Ptolemy's in the sense that it used the ideas of epicycles, made the revolutionary suggestion that the Earth was not the centre of the Universe. Copernicus begins his book by saying that the Sun lies at the centre of the universe – although as the book progresses he contradicts this, perhaps because he was aware that this suggestion would not be well received. You will have noticed the fact that the date of publication of the book is the same as the date of his death. This is no coincidence: he waited thirty years before publishing it, and presumably only finally let it be published when he knew he was beyond religious persecution. Copernicus suggested that the apparent circular motion of the stars was caused not by the rotation of crystal spheres, but by the rotation of the Earth itself, and that the occasional retrograde motion of the planets was caused by the fact that they orbited the sun at different rates, rather than the Earth. This downgraded the Earth to being just another planet. It shifted humanity from centre-stage.

Because his model was so powerful, Copernicus rejected the notion of it as being merely a description. Instead, he adopted the view that it was *true*: that his model was a description of reality. Perhaps it is this that is the real breakthrough in the book, rather than the fact that it moved the Earth from the centre of the universe.

To make such a claim must have taken immense courage, and perhaps even a touch of arrogance. Copernicus was saying that the authorities, going all the way back to the ancient Greeks, were wrong, and that this model, *his* model, was not merely a description of our *perceptions* of things, but was in fact a description of the way things worked, of *reality* itself. He was saying that his model in some sense encapsulated God's creation: he had captured God's thoughts.

Of course, this summary is somewhat distorted. Copernicus existed at a time when others were also moving in the direction of a realist interpretation of mathematical models, although Robert Westman has suggested that only 10 others regarded Copernicus's models as a representation of the truth before 1600 (Westman, 1980).

Another nail in the coffin of Aristotelian perfection was the demonstration, based on observational data, by Tycho Brahe (1546–1601) that comets were not, as had previously been thought, events which occurred in the atmosphere, but in fact took place much further away from the earth. So far away, in fact, that their orbits must pass through Aristotle's crystal spheres. Imagine the shattering consequence of that! It meant that, instead of being embedded in some physical substance, the rotation of which led to the planets and stars moving, they now moved in their own right, with nothing holding them up. I can feel the very foundations of the conceptual universe shifting as I write about this change: with nothing to hold the stars up, why don't they fall on us?

Johannes Kepler (1571–1630) moved things even further along the route towards mathematical models as models of reality with his description, based on an analysis of the wealth of precise observations collected by Tycho Brahe. The unprecedented accuracy of Brahe's data enabled Kepler to spot a slight difference between the observed position of Mars and the position as predicted by Copernicus's model. From this, Kepler deduced that the planets *did not follow circles at all*, but orbited in elliptical paths with the sun at one focus. The accuracy of this model lent very strong support to the notion that the Earth was not at the centre of things. Kepler described his ideas in two books, *Astronomia nova*, published in 1609 (Kepler, 1992), and *Harmonices mundi*, published in 1619 (Kepler, 1997). In these books, he presented the summary of his theory in terms of three laws, which still (four hundred years on) bear his name. Kepler also suggested that some kind of force (by analogy with magnetism, he suggested) between the planets and the sun could be responsible for the elliptical orbits. The role of mathematics in science was moving beyond being mere convenient description. It was becoming a representation of reality.

Kepler's ideas were based on the data collected by Tycho Brahe, but another revolution was brewing that was to provide another source of data so rich that science would soon sweep the ancient astronomical theories away for good. In 1608, a Dutchman called Hans Lippershey invented a device which would extend human vision dramatically – the telescope. I have already remarked that the introduction of this new instrument met with some resistance, but within two years of its invention, Galileo Galilei had built a series of such devices, grinding his own lenses, and producing one which could magnify up to 30 times, sufficiently powerful to see the moons of Jupiter. In particular, he noticed that these moons formed a model of the entire solar system in miniature, with smaller bodies orbiting a massive central body. And the observational hammer of data drives its nail into the Aristotelian coffin: the paths of the smaller bodies were ellipses.

Galileo Galilei (1564–1642) is probably the best known figure in this story. The Catholic Church did not officially pronounce on the Copernican description after its publication in the year of Copernicus's death in 1543, but years later, in 1616, it suspended its approval of the book pending 'correction'. It clearly did not regard the book as a great threat to its doctrines. The Catholic Church, in the figure of Cardinal Robert Bellarmine, viewed Copernicus's theory in just the way described above – as a convenient calculating tool which yielded good predictions of the positions of astronomical bodies – but not as a proposed representation of reality. That would have been too presumptuous. Unfortunately, Galileo, in a letter to the Grand Duchess Christina of Lorraine, took the opposite view. He said 'I hold that the Sun is located at the centre of the revolutions of the heavenly orbs and does not change place, and that the Earth rotates on itself and moves around it.' He based his claim firmly on data, referring to 'physical effects' and 'astronomical discoveries'. Interestingly enough, he did not base his ideas solely on his telescopic observations, but also on his theory of falling bodies. Nevertheless, he clearly took the view that the data, the observations of physical phenomena, should dominate, and that they dictated the way that nature must be, so that his theory was not merely a 'convenient calculating tool'.

Despite the disagreement between himself and the Church, Galileo clearly believed he was not at risk, even going so far as to dedicate his book *Il Saggiatore* to the Pope, writing

> Philosophy is written in this grand book, the universe, which stands continually open to our gaze. But the book cannot be understood unless one first learns to comprehend the language and read the characters in which it is written. It is written in the language of mathematics, and its characters are triangles, circles, and other geometric figures without which it is humanly impossible to understand a single word of it; without these one is wandering in a dark labyrinth.
>
> (Galileo Galilei, *Letter to the Grand* Duchess, 1615)

The grip of mathematics as a way of describing the universe was tightening.

Galileo had several papal audiences with Pope Urban VIII, and continued to believe the Church would not regard his theories as contradicting their doctrine, but when his book *Dialogue Concerning the Two Chief Systems of the World – Ptolemaic and Copernican* was published in 1632, the Church immediately banned it. Galileo, now 69, was summoned before the inquisition, where he was charged with breaching the conditions laid down in 1616 by promoting the notion that the Earth moved. At his sentencing on 22 June 1633, however, it was agreed that if he recanted he would not have the full force of the penalty imposed upon him. Confronted by instruments of torture, what could this frail old man do but recant? As a result, he was sentenced to life imprisonment (though it was a fairly minor form, really amounting to house arrest). The story has it that, as he was rising from his knees after conceding that the Copernican theory was wrong, he whispered '*eppur si muove*' – '*but it does move*'. While imprisoned, he worked on *Discourses and Mathematical Demonstrations Concerning the Two New Sciences*, which was smuggled to Leiden, where it was published. He died while imprisoned.

Galileo also played a wider role in the shift towards mathematical models as representing reality. Earlier in his career, he recognized that, unhindered by air resistance, all bodies would fall with the same acceleration, in contrast to Aristotle's position that the speed of fall was proportional to the body's weight. This is the substance behind the famous, but probably apocryphal, story of him dropping two different weights from the top of the tower of Pisa, and demonstrating that they hit the ground together: a perfect illustration of collecting data. Likewise, he was able to combine ideas of the downward motion of a falling body with its horizontal motion due to initial impetus or the motion derived from the earth's rotation to yield an elegant overall summary of the motion of bodies. Opponents of the rotating Earth theory had argued that, if the Earth rotated, a body dropped from the top of a tall tower would fall behind the foot of the tower: as this did not happen, the Earth did not rotate. But Galileo's theories also led to the observed result, despite the rotation. His theories also predicted the parabolic path we see in a thrown ball.

Galileo also explained various other mechanical phenomena, such as why a body like the earth could continue to orbit the sun without any apparent external force (which Aristotle would have required), why a ball rolled up an inclined plane will slow down, while one rolled down will speed up and why a ball in a horizontal plane will remain where it is – or, if given an initial motion will continue with that motion until interfered with. The moon moved about the earth in just such a way. Apart, that is, from the point that the moon's orbit is not circular – it moves away from and towards the earth, tracing an ellipse, as it moves. The ground was set for Newton's great work.

Gradually, then, over the sixteenth and seventeenth centuries, acceptance of authoritative descriptions of how things worked, such as those given by Aristotle, began to lose ground in the face of contradictory evidence from the real world. An Aristotelian view which is, in some arbitrary sense 'perfect', loses its relevance when an alternative description, based on observations of data, yields better predictions. It is hardly surprising then that what is regarded as a model of reality shifts from the idealized and inadequate Aristotelian view to the mathematical model. The title Newton chose for the great work in which he showed, amongst many other things, that the force which led to the planets orbiting was the same as that which made an apple fall, was the *Mathematical Principles of Natural Philosophy*. Mathematics had become recognized as a model of the way things really worked, and not merely a convenient description. Natural philosophy had graduated and become something else: something called science.

Sciences such as astronomy and physics rest very heavily on mathematics – they have mathematics at their very core. In fact, one might say that they have a particular kind of mathematics at their very core. Some have taken this as defining science: the Nobel Prize winning physicist Lord Rutherford once said 'all science is either physics or stamp collecting', contrasting the aim of physics to find simple laws that describe how the universe functioned with the naturalists' detailed study of endless different kinds of plants and creatures. The truth is that there are different kinds of science. As contrasts to physics, we might cite biology and the social sciences. Classical biology has indeed been concerned with collecting vast amounts of data, describing the various types and families of animals and plants, but more advanced modern biology does not try merely to describe these. Instead it tries to relate and understand them. Here mathematics, and especially statistics, plays a very important role. In the past, biological families were described by their observable shapes and structures – their *morphological* characteristics: chimpanzees and humankind are primates because of their obvious similar characteristics, such as flexible hands. However, with the advent of molecular biology it became possible to look at the very genes describing how such animals were put together. This led to the use of advanced mathematical tools to calculate distances between the genes, and to defining families in terms of how close the animals were in an evolutionary sense. For example, one could ask whether the evolutionary split that led to these two organisms developing in different directions occurred in the near or distant past. These developments in *genomics* are exemplified by the human genome project, and more recently, by the application of similar tools to protein structures in *proteomics*. In all these situations, mathematics and data analysis play an absolutely fundamental role. The data sets are so large and complicated that, without sophisticated techniques to tease them apart, there would be no hope at all of understanding them. Modern biological science could not exist without mathematics – not because its core structures take the form of differential equations as in physics, but because extracting meaningful patterns from vast data sets requires careful and sophisticated analysis of the data, using the sort of tools described in Chapter 5.

The social and behavioural sciences are different again. Whereas physics deals with basic entities such as electrons, *all of which are identical*, these sciences deal with basic entities such as people, *all of which are different*. One might thus expect these sciences to be considerably tougher than physics, and this, some have argued, explains why the extraordinary advances in physics, and the consequent progress in physical technology which forms the backdrop of the modern world, have not been matched by corresponding advances and applications in economics and sociology. We still have fairly fundamental debates and arguments about human behaviour but we do not have debates or arguments about the likely consequences of cutting the temperature when manufacturing plastic. Despite all this, economics especially has become a mathematical science: that particularly mathematical part of it, called econometrics, seems to be modelled largely on the physical sciences. The Nobel prize-winning economist George Stigler has remarked that it is necessary to identify the 'uniformities in the subject matter' in order to gain useful understanding in economics, saying that 'in the present early stage of economic study, the economist as scientist must be largely occupied with the isolation of these uniformities in his subject matter ... until we possess many uniformities, we cannot erect broad analytical systems which are likely to be illuminating in the area where uniformities have not yet been isolated' (Stigler, 1949). This sounds suspiciously like physics.

There are also other important differences between the physical sciences, on the one hand, and the human, social, or behavioural sciences on the other. Physical objects, of course, do not see you taking the measurement and certainly do not then decide to behave in a deliberately contrary way. An electron does not maliciously decide to fly off in the wrong direction, but people may. Moreover, people in different cultures may respond differently to the same stimulus – whereas, again, electrons behave in the same way wherever they are. Finally, in carrying out experiments on electrons, one does not have to be worried about ethical or moral issues: electrons feel no pain. The same cannot be said of experiments involving humans or animals. So there are differences between sciences, and in the extent and way they use mathematics.

Perhaps a fitting way to conclude this section is to quote a passage written by John Arbuthnot in 1692 in the preface to a translation of a work by Christiaan Huygens (Huggers, 1920). The work was entitled *De Ratiociniis in Ludo Alae* (*Calculating in Games of Chance*) and was originally written in 1657, when the basic concepts of probability were first being formalized. Arbuthnot wrote: 'There are very few things which we know, which are not capable of being reduc'd to a Mathematical Reasoning, and when they cannot, it's a sign our Knowledge of them is very small and confus'd.'

Science and religions

Science and the many religions are often painted as antagonists, but this is a gross oversimplification. This perception is coloured by historical episodes such as the imprisonment of Galileo for promoting the heliocentric view of the solar system, described above, but this happened 400 years ago. Both science and the interpretation of (at least some) religions have evolved substantially during that time.

Religions originally developed as a way for people to understand the randomness, the arbitrariness, and the general awesomeness and unpredictability of the world. That is, they provided a framework through which to interpret the world. They also provided sets of moral and ethical guidelines, telling people how they should lead their lives, and what was 'right' in the context of their time and their society. This is very evident in many of the traditions which linger on today (e.g. concerning what may be eaten and how it should be prepared, and what clothes should be worn), bearing little relationship to modern life, but having a clear basis in the daily life of centuries ago, when they served to protect people from illness and mishap. For example, such traditions provided effective rules in the context of a lack of understanding about the mechanisms causing illness arising from contaminated food. Of course, the notion of 'rules for how one should lead one's life' becomes intertwined with how rulers wished their subjects to lead their lives. Religion and the state have often been closely entangled.

As we have seen, such frameworks, constructed in a way largely divorced from evidence, are intrinsically shaky. In his work *An Inquiry Concerning Human Understanding* (1748), the philosopher David Hume (1711–1776) wrote 'If we take in our hand any volume of divinity or school metaphysics, for instance, let us ask, Does it contain any abstract reasoning concerning quantity or number? No. Does it contain any experimental reasoning concerning matter of fact and existence? No. Commit it then to the flames, for it can contain nothing but sophistry and illusion' (Hume, 1999). And recall again Francis Bacon's warning against giving out 'a dream of our own imagination for a pattern of the world'. Furthermore, there is no end to such dreams. We can easily invent alternative dreams, and human history is littered with them. Think of the Norse Gods, the Greek Pantheon, the Roman gods, the Sun gods of Egypt, the various gods of the Aztecs, as well as the many different religions that have held that there is one true (all rather different) God.

The way out of this tangled maze of beliefs, all held with an intensity that has led to untold numbers of deaths and an unimaginable degree of suffering, is to look at the evidence, the data. Of course, as we have seen in the evolution from magic to science, although revolutionary, the shift to an evidence-based way of understanding nature was not a sharp schism. The early development of scientific ideas, and the first scientists, existed in a world permeated by religion, with religion providing the infrastructure on which interpretation took place. In the unlikely event that someone was able to step completely out of this framework, their ideas would not have been understood, and they would have been ignored or worse – think of Galileo and the fate of others classed as heretics or whatever was contrary to the religious context of the time and place in question. The change had to be gradual, even if, like any other change, it was painful.

A study of the life of many of the protoscientists such as John Dee (1527–1608) or Isaac Newton (1642–1727) shows clearly this interweaving of the existing framework with that provided by the new insights arising from experimental or observational evidence. It also often shows the intellectual pain that such transitions and attempts to understand can cause. Indeed, one can go back even further, to such people as Robert Grosseteste (c.1168–1253) and Roger Bacon (c.1220–1292). We have already met Roger Bacon, and Robert Grosseteste was one of his teachers. Grosseteste lectured in theology to the Franciscan school in Oxford from 1229 to 1235, when he was appointed Bishop of Lincoln, addressing the papal congregation at Lyon in 1250. His 'scientific' work was concerned with geometry and optics, as well as astronomy, and he believed in practical experiments to test theories, and used mirrors and lenses. Perhaps his most famous work (of around 120 which he wrote, many on theology, philosophy and church leadership, as well as science) is *De Luce* ('Concerning Light'), in which he argued that light is the foundation of matter (hence the importance of God's command 'let there be light') and understanding (hence our sayings 'I saw the light' and 'light dawned').

We see from the above that, to a large extent, science took over from religion. Science, a mode of reasoning and understanding gained from study of the evidence, led to an alternative way of comprehending nature. In this sense, one can certainly think of it as a competitor to religion. The aim of science is to discover how the universe, in the broadest sense, works, and one might paraphrase this as aiming to discover the 'mind of God'. Religion is also concerned with knowing God, but there are key differences. Science was not merely yet another alternative to the many religions that had been coined; in particular, science did not work at the same sort of level. Science, based on Baconian principles of observation and experiment, was another kind of animal altogether.

First, because of its foundation in observation, science had predictive power. When A happens, B will happen. If you do X, then Y will follow. This understanding led to technological advances. Technology is the offspring of science concerned with using scientific discoveries. It is modern technology, based on scientific understanding, which gives us the automobile, the aeroplane, the computer, the skyscraper, and so on. Indeed, the very food we eat and the clothes we wear are produced by a highly complex process of many steps using advanced technologies. Religion has no such predictive, *manipulative*, power. Of course, some people believe in the possibility of intercession by the gods or a God as a result of prayer or sacrifice, but the evidence, the data, showing that this works does not exist.

Second, science (and good scientists) explicitly acknowledges its fallibility. The very essence of science is the notion of conjecture and test. If a theory is untestable, it lies outside the realm of science. As we have already seen, testability, refutability, falsifiability are what makes science science. It is this property which distinguishes science from Bacon's 'dreams of our own imagination'.

Third, as a consequence of this principle of falsifiability, science is not static. It evolves as new, more powerful, accurate, or broader theories replace earlier ones, in the sense that they can explain the data better. The acknowledgment that scientific understanding is a process of constant change is very different from the religious perspective.

Fourth, because of this evolution, science is cumulative: its predictive power, the breadth of the understanding it gives, is constantly increasing.

Fifth, although there are many religions, in a particular sense there is only one science. Obviously there are different scientific disciplines, in the sense that biology, chemistry, physics, palaeontology, archaeology, rheology, etc. are all different, but the scientific process, the gradual accretion of evidence, the construction of theories which are compared with the evidence and so on, is essentially the same. Moreover, although different biologists may hold different theories, if these theories can be properly called scientific, then it will be possible to compare them. It will be possible to collect data to choose between them in terms of their power to predict how the world behaves.

Science, then, is predictive, fallible, changing, growing and unified in a way which religions are not.

There are also other differences, of course. I mentioned the role of religions in providing moral and ethical guidelines: in telling people what was 'right', ranging from the belief that one should never lie, to the religiously motivated suicide bomber. This role of religion is entirely outside science. Science seeks to uncover what makes things tick, not tell us how we should behave. In the interests of balance, perhaps I should add that, of course, religion is not the only source of ethical and moral guidelines. The American Humanist Association, for example, defines *Humanism* as 'a progressive lifestance that, without supernaturalism, affirms our ability and responsibility to lead meaningful, ethical lives capable of adding to the greater good of humanity'.

CHAPTER FOUR

Big Brother's Eyes

There was of course no way of knowing whether you were being watched at any given moment ... It was even conceivable that they watched everybody all the time.
George Orwell, *1984*

Your every movement, your every action

Imagine a society in which anyone who committed a crime was *guaranteed* to be caught. Sounds perfect? Levels of crime would plummet. Quality of life would rocket. We could leave our front doors open without fear of intruders walking in and stealing our jewellery or laptops. We need have no fear of being mugged or raped as we walked along the streets.

This would not be quite true, of course. There are always maniacs or terrorists not bound by the usual fear of being detected, caught and punished, but it would certainly go a long way towards improving everyday life.

If this sounds wonderful, then be aware that such a society is certainly within our reach. In fact, every day we see signs of it beginning all around us. It is achieved by collecting data about each of us, about our activities and about our whereabouts as we go through the day. The UK, for example, has the highest per capita number of closed circuit TV surveillance cameras in the world, with an estimated number of around four million cameras, but other countries are not far behind. These cameras silently observe us as we walk or drive to work, as we enter shops or banks, as we buy tickets for public transport, and as we travel on trains and buses. Certainly the quality of many of the pictures leaves much to be desired, but that's easily solved by installing superior TV systems, and that will happen soon enough as their cost continues to fall with advances in electronics technology. By monitoring us in this way, the cameras discourage gangs of youths from causing disturbances in the streets, they mean that only the stupidest of crooks tries to rob a bank (although there are some stupid ones out there – such as the guy who forgot to cut eyeholes in the wool hat he rolled down over his face to protect him from the cameras), and they silently observe the man lifting your purse from your bag while his companion distracts you.

More sophisticated systems – and things are becoming more sophisticated day by day – even go so far as to read your car number plate and recognize the faces of terrorist suspects or football hooligans in crowds.

The idea of being observed in this way is not a new one. Writing in 1787, Jeremy Bentham described the *Panopticon*. This was a design for a prison in which a single central tower could observe rings of prison cells around it, through one-way glass, leading to 'the illusion of constant surveillance' (Bentham, 1995). The philosopher Michel Foucault has commented that 'modern society increasingly functions like a super Panopticon' with individual behaviour being controlled by the fear of being under constant observation, and one can see that the closed circuit TV cameras make this a painful reality. (Slobogin, 2002).

However, the cameras are just one aspect of surveillance. In addition to cameras, each day individual records of our purchases in supermarkets, of our bank transactions, of our bus and train journeys, and so on, are recorded and subsequently pored over and analysed in minute detail by people who are expert at squeezing information from these records. These data can be used to determine whether someone was where they said they were on a particular Thursday night last year. They can be used to determine whether you are spending more than you appear to earn. They can be used to determine how you are likely to behave in the future, and to detect departures from that behaviour. So, for example, if you normally fill up your gas tank every Sunday night, in preparation for the week ahead, but then make three gas station purchases within three hours on a Saturday afternoon, a flag can be raised. Could it be that your credit card has been stolen? (Or could it be that your three sons have driven to a family gathering, and you have kindly filled their tanks for their return journeys?)

Radio Frequency Identification (RFID) systems are becoming increasingly common for controlling stock in shops or warehouses. They enable the identity of objects to be verified automatically without line of sight contact, but such systems can be used to locate and track people. Think of the system which will scan US passports from a distance. At one level, this is a tremendous help to the individual – removing all those interminable queues at US immigration. At another level, it raises the spectre of Big Brother, watching our every move. Parents of children at Brittan Elementary School in California experienced this first hand, when an RFID system for monitoring student movement was introduced. The nominal aim of the system was to track attendance, so that it would quickly be realized if a student was missing, and to detect trespassers, both objectives which are obviously highly desirable. Rather to the authorities' surprise, however, parents strongly objected, and some refused to allow their children to carry the ID cards containing the RFID tags. Things then escalated, with the school threatening disciplinary action if people did not take part. This in turn precipitated the involvement of the American Civil Liberties Union of Northern California.

Perhaps the motives behind the proposed school RFID student tracking system are not quite as clear-cut as they seem. School funding in California is based on attendance, so that accurate counts are important to the school, providing an audit trail. Providing an attendance audit trail is of only indirect benefit to the individuals being monitored. It benefits the group, or the overall school, rather than any specific individual being monitored. While the school tried to sell the merit of the system as able to monitor individual students, perhaps the higher level interests were the real aim. This raises the question of how much intrusion into individual behaviour is legitimate when broader group interests are at stake.

At least in the Brittan Elementary School case, people knew they were being monitored. Consider using a credit card in a shop to purchase an RFID labelled item. Not only can the purchaser carrying the item be tracked, using the RFID signal, but they can also be uniquely linked to that item via their credit card record: all unknown to the customer. Very useful if the credit card has been stolen or cloned – just picture the surprised look on the crook's face as he opens his door to the police.

In short, such data can be used to detect crime, to prevent it, and to identify those responsible when it happens, but it comes at a price. To detect the small percentage of individuals committing a crime, everyone must be monitored, and not everyone likes the idea that their movements are being watched, that the patterns of things they buy are being studied, or that the identity of the people they talk to is known.

In 1787, George Washington wrote that it is necessary for individuals 'to give up a share of liberty to preserve the rest', and in the same year, Alexander Hamilton wrote that 'the violent destruction of life and property incident to war, the continual effort and alarm attendant on a state of continual danger, will compel nations the most attached to liberty to resort for repose and security to institutions which have a tendency to destroy their civil and political rights. To be more safe, they at length become willing to run the risk of being less free'. On the other hand, Benjamin Franklin wrote 'they that can give up essential liberty to purchase a little temporary safety deserve neither liberty nor safety'(Franklin, 1963). It's all a question of balance, between, on the one hand, protecting society from crime, terrorism, or simply accident, and on the other, monitoring the behaviour of individuals to the extent that their privacy is invaded. This chapter looks at some of the data technologies underlying these issues. Later chapters look at the manipulation of data to extract meaning and value, as well as its manipulation for dishonest purposes.

This notion of striking a proper balance between respecting and invading personal privacy is complicated, because notions of privacy vary between societies. Moreover, as far as crime is concerned, its very meaning is a societal construct, depending on the country, the social mores, and the time or era in which the crime occurred. Some things contravene social conventions without being criminal. Smacking children, dropping litter, and perhaps even wearing revealing clothes in public are examples of things which hover at the edges of the criminal, with their status depending on where and when they are carried out. Of course, as our world continues its slide towards political correctness, who knows what might be criminal next year? Dropping chewing gum is a crime in some countries (in Singapore, for example, though perhaps not as Procrustean as it seems, since dropped gum jammed the automatic doors of the underground trains), but not in others.

The fact that governments need detailed personal data to make sensible decisions as to how to run the country, and about what is best for us now and in the future, means that they are also under legal and ethical obligations to preserve the confidentiality and privacy of the individuals providing the data. Without public confidence that confidentiality and privacy will be preserved, survey and census response rates will fall, which will lead to less accurate data, which in turn will lead to less effective government. It is in the government's interest to ensure that people are confident that their privacy will be preserved, and that the data will be used in a proper and responsible way. This does not apply merely to governments, of course. If a corporation is to produce a product or service that matches your wishes, and to make you feel special because it is giving you what you want, then it needs to know what your wishes and wants are.

In an attempt to strike a suitable balance, one might formulate a basic principle of privacy along the lines that '*personal data should be used only for the purpose for which it was collected, unless explicit permission is given*'. Many governments have expanded this into underlying guidelines such as:

- data should be obtained lawfully and fairly.
- data should be relevant to the purpose for which it is to be used, as well as being accurate, timely and complete.

(Extraordinarily, in March 2003, the US Justice Department exempted the FBI's National Crime Information Center from the US Privacy Act's requirement that data be relevant, accurate, timely and complete!)

- the purpose for which the data are to be used must be specified, and data will be destroyed if they are no longer relevant to that purpose.
- personal data should be used only for the purpose for which it was collected, unless the explicit permission of the person described by the data is obtained.
- safeguards against loss, corruption, destruction, or misuse must be in place.
- information must be available about the collection, security, and use of personal data.
- the subject of data has a right to inspect and challenge the data relating to them.
- an accountable officer should be responsible for ensuring compliance with the above.

In addition to regulations derived from the above, there are also sometimes further regulations relating to particularly sensitive data, such as that applying to children, health, race, religion, or finance.

The balance between personal privacy and public good is an aspect of the balance between the needs of individuals and the needs of groups. These may not be aligned, and what is of interest to one may not be of interest to the other. At a simple level, this is obvious. Aggregate data on crime matter to the government, so it can decide what size of police force is needed, but individual crime matters to you, so you can decide whether it is safe to walk in the street. Aggregate data on the number of children below 16 years old in a town matters to the local authorities, who have to plan how many schools and schoolteachers they need and must pay for. Individual data on whether you are below 16 years old matters to you since it affects what movies you can see and whether you can buy alcoholic beverages.

While collecting detailed data on individuals may threaten their individual privacy, it is very clear that it has the potential for immense good for the group. Examples abound.

- In epidemiology, monitoring how and when people fall ill can lead to the early detection of small local clusters of illness, thereby preventing the spread of diseases such as SARS from causing major epidemics. At the time of writing, the world appears to be at risk from avian flu. An awareness of when and where people are falling ill could save tens of millions of lives.
- In post-marketing surveillance of pharmaceuticals, studying who suffers from what side effects of what prescribed medication can lead to the early detection of serious problems, and avoid a great deal of suffering. In fact, barely a week goes by without the popular press reporting suspicious data arising from some widely used medication. At the time of writing, the pharmaceutical giant Merck has recently announced a voluntary worldwide withdrawal of Vioxx. Vioxx is a COX-2 selective non-steroidal anti-inflammatory (NSAID) prescription medicine used for pain, such as that arising from arthritis. Monitoring large numbers of users has suggested that Vioxx may cause an increased risk of heart attack and strokes if used over a long period.
- Monitoring telephone calls (not the conversations, but the time of the call and the location of the caller and recipient) is clearly a potentially gross invasion of privacy, but this sort of exercise enabled the security services to track one of the failed 21 July 2005 London alleged terrorist bombers from London to Paris to Milan, and finally to Rome, where he was arrested.
- Nowadays, genetic fingerprints can be obtained from microscopic traces of organic matter – from, for example, the saliva left on a used cigarette butt. This needs to be matched to a database of genetic fingerprints – which requires everyone to provide such a record.

- You would naturally be very unhappy if you discovered someone had been using your credit card or had taken all your money from your bank account. To prevent this, the bank must be certain that it is really you making the transactions, and not some imposter. This requires the bank to ask for verification of your identity prior to each transaction. It means the bank has to keep data on you so it can make such verifications. It also means the bank has to monitor your transaction behaviour so it can detect if things have gone awry: that the unusual sequence of transactions is really you doing something out of the ordinary, and not a thief spending as much as he can before the card is stopped.

Often a fairly deep level of personal detail must be recorded in order to fulfil aims of the kind outlined above. It would seldom be sufficient to try to detect potential terrorists amongst air passengers solely on the basis of their names – and attempts to do just this have led to embarrassing and painful consequences. Senator Edward M. Kennedy was prevented from boarding his US Airways Washington to Boston flight when he was mistakenly matched to someone on a list of suspicious persons. Later, he was also automatically flagged for observation by a system which looks for suspicious behaviour, such as buying a one-way ticket.

Senator Kennedy's misadventures arose as a consequence of increased data collection on individuals, a result of the World Trade Center atrocity. In 2002, the US Defense Advanced Projects Agency (DARPA) launched its Total Information Awareness (TIA) program, with the objective of mining data describing people's movements, transactions, and interactions in order to detect potential terrorists before they could act. The Director of the US Information Awareness Office described the need to become 'much more efficient and more clever in the ways we find new sources of data, mine information from the new and old, generate information, make it available for analysis, convert it to knowledge, and create actionable options'. The TIA programme stimulated considerable anxiety in Congress and the press, where it was seen as risking an invasion of privacy. William Safire wrote a column in the *New York Times* on 24 November 2002 raising some concerns. He wrote

> Every purchase you make with a credit card, every magazine subscription you buy and medical prescription you fill, every Web site you visit and email you send or receive, every academic grade you receive, every bank deposit you make, every trip you book and every event you attend – all those transactions and communications will go into what the Defense Department describes as 'a virtual, centralized grand database.'

> To this computerized dossier on your private life from commercial sources, add every piece of information that government has about you – passport application, driver's license and bridge toll records, judicial and divorce records, complaints from nosy neighbours to the FBI, your lifetime paper trail plus the latest hidden camera surveillance – and you have the supersnoop's dream: a 'Total Information Awareness' about every US citizen.

In May 2003, the programmes name was changed to the Terrorist Information Awareness Program (keeping the same acronym), and in February 2003, the US Secretary of Defense, Donald Rumsfeld, created the *Technology and Privacy Advisory Committee* (TAPAC) to look into the use of 'advanced information technologies to identify terrorists before they act', and in particular, of course, the relationship with US privacy laws. In fact, the report also considers non-US privacy protections and principles – commenting, for example, that 'the *Wall Street Journal* reports that these [European Union] laws are so restrictive that they prohibit a business from making its corporate telephone directory accessible from non-European countries, if it contains office telephone numbers of individual employees'. In September 2003, Congress stopped the programme's funding, apart from funds for the 'processing, analysis, and collaboration tools for counterterrorism foreign intelligence'.

Other similar programmes exist. There are many examples.

- The US Department of Homeland Security prescreens airline passengers, comparing names with private and public databases to assess the level of risk they pose.

- The US Financial Crimes Enforcement Network monitors financial transactions with a view to detecting and preventing money laundering.
- There are increasingly elaborate linking systems permitting law enforcement records to link into other government databases.

The Report of TAPAC published in March 2004 reached four main conclusions:

1. TIA was a flawed effort to achieve worthwhile ends, flawed because of its apparent insensitivity to privacy issues, and the manner in which it was presented to the public.
2. Data mining is a vital tool in the fight against terrorism, but when used in connection with personal data concerning US persons, data mining can present significant privacy issues.
3. In developing and using data mining tools, the government can and must protect privacy.
4. Existing legal requirements applicable to the government's many data mining programs are numerous, but disjointed and often outdated, and as a result may compromise the protection of privacy, public confidence, and the nation's ability to craft effective and lawful responses to terrorism.

The report went on to say 'We believe it is possible to use information technologies to protect national security without compromising the privacy of US persons', and it made a series of twelve administrative recommendations for overseeing such technologies.

The danger of errors in databases makes the situation worse. As Chapter 6 illustrates, although we might like to believe otherwise, it is highly likely that our records contain many errors, whoever keeps them and whatever they are records of. The TAPAC report drew attention to the danger that individuals may be profiled on the basis of inaccurate or out of date information, and this is the case at even the simplest of levels: we have already seen that one of the central guidelines of privacy is that data should be relevant, accurate, timely and complete. Of course, one may not discover that this is not the case until it is too late: until the bank loan has been turned down or the mortgage declined.

One is not reassured by the report that US border guards failed *100 per cent of the time* to detect counterfeit identity documents being used by General Accounting Office agents trying to enter the US illegally as tests.

One of the great difficulties in detecting terrorists arises from the fact that there are so few of them. A similar issue arises in detecting banking fraud, where the legitimate transactions greatly exceed the fraudulent ones. What the great imbalance between the number of terrorists and the number of ordinary citizens means is that, inevitably, there will be many 'false positives' – people flagged as potential terrorists who are completely innocent. This is unavoidable.

As Senator Kennedy can tell you, and as many who have had their credit card transactions blocked will know, mistakes will occur. There is no such thing as a system which will never invade anyone's privacy while always optimizing the public good. It is all a question of balance.

Geodemographics

Geodemographics is the study of how people behave and what they are like in relation to where they live. We are all familiar with its basic ideas from the newspapers. During the 2004 US Presidential election, maps of the US with the 48 contiguous states coloured red or blue according to whether their population voted Republican or Democrat showed a remarkable pattern, with states to the west and north-east coloured blue and all the intermediate states coloured red. It was immediately clear from this map which states voted which way: but in fact, the map was rather misleading, because it seemed to suggest that many more people voted Republican than Democrat, whereas as we know, the population was almost evenly divided. This misleading impression arose from the fact that the coastal areas are more densely populated than the inland areas. Distorted versions of the map, in which a state's area on the paper represents the size of the population in that state, showed a much more equal distribution of numbers of votes.

Such maps summarize, in simple and visually accessible formats, data on how people in different geographical areas behave – in this case, in terms of voting behaviour. But this is only the tip of the geodemographic iceberg. Voting is just one aspect of behaviour. If we are planning to establish a new retail outlet, we must plan carefully where to locate it. There is no point in putting a shop specializing in clothes for newborn babies in an area predominantly populated by young singletons, or in promoting young singleton lifestyle choices in an area predominantly occupied by people of retirement age. Rather, we want to put our babywear shop in an area where there are likely to be many young families, or in a spot which is easily access-ible from such areas. From an another perspective, we might study the characteristics of a population to decide what services to put there: should we organize a bus service, for example? Can a market niche be exploited in a particular area: perhaps a coffee shop, an antique shop, or a delicatessen? To make such decisions and to answer such questions, we need detailed data on the local population and its behaviour.

We would also like to know whether it is worthwhile advertising a particular service in an area. Would a poster placed on a hoarding at a particular location be a waste of resources? Mass junk mailing can be spectacularly unsuccessful, and even counterproductive. People rapidly learn to bin the mass of leaflets falling through their doors – and each leaflet costs money to send – but they also become irritated by them. Spam email is even worse, at least partly because of their huge numbers, each requiring a small slice of time to recog-nize and delete. (Spam filters are now highly effective at ridding us of this curse, but, of course, they will also filter out any advertisement we might be interested in. I carry out research in the retail financial services, so have to be very careful that my spam filters do not prevent perfectly legitimate emails about mortgages or bank loans from research colleagues from getting through to me.) What is needed in place of both of these sorts of operations is targeted mailing (or emailing), which would mean that when an advertisement or promotion does reach you, it is much more likely to be of interest to you. However, to achieve such targeting, one needs to have a great deal of data about the population so one can separate out the targets from the mass of people who would not be interested.

One important way to achieve this is through population segmentation, that is, by dividing the population up into distinct groups according to their behaviour. Behaviour will be measured by surveys asking relevant questions and from linking the responses to other sources of information, such as electoral rolls and area data derived from census records. Then, people with similar behaviour profiles will be grouped together into a small number of groups. Quite what sort of segmentation one will aim for will depend on what sort of behaviour one is targeting. A segmentation for financial behaviour may be very different from one for clothes and fashion.

Examples of general lifestyle and behaviour segmentations are ACORN (A Classification of Residential Neighbourhoods) in the UK and PRIZM (Potential Rating Index for Zip Markets) in the US. An example of a financial behaviour segmentation was the FRuitS system, which analysed behaviour in terms of life-stage, financial status and financial product portfolio, dividing a population up into eight classes. The propensity of members of different classes to take new financial products differed substantially between the classes (you can probably guess the sorts of characteristics of some of the classes from their names: the lemons, for example, or the plums). An analysis cross-referencing the FRuitS class against other (more accessible) characteristics enabled researchers to allocate new people to a FRuitS class on the basis of these other characteristics, so that entire populations could be assigned to a class without having to survey an entire population.

Such segmentations are commonplace, and one often reads examples of new ones being generated by geodemographics companies (often, one suspects, to keep the name of the corporation in the public eye, as much as for any genuine value which is associated with the new classification).

Geographic segmentation is an attempt by governments and commercial operations to impose some sort of taxonomy or classification onto the immense diversity of people and their behaviour, so that sensible administrative and marketing decisions can be made. So, while one might object to being lumped together with others whom one regards as quite different, it is less easy to object on the grounds of invasion of privacy. Of course, such lumping together does necessarily imply that everyone who lives in a certain geographical area actually has the same profile of characteristics. For example, there is no reason why a group of students should not rent a house in an area predominantly populated by elderly pensioners. This was very apparent from the coloured maps of the outcome of the 2004 US presidential election. If, instead of colouring *states* according to whom the majority of their voters voted for, one colours *counties*, a very different, and much less uniform picture emerges. While there is still a greater density of blue areas around the coasts, they are interspersed with red. And the middle of the country has quite a few blue patches scattered across it. In fact, even these maps give a rather distorted impression, since they do not distinguish between a county in which a slim majority voted Republican and a county in which a large majority voted Republican. Both are coloured red.

Such segmentation systems, such ways of imposing structure, are likely to become more important in the future, not least because of the major cultural changes we are being exposed to. Not so long ago, there were only a few TV channels to choose from, so that people had a similar cultural experience – and, for example, one could talk at work about a TV programme one might reasonably have expected one's colleagues also to have seen, but this is no longer the case. Not only are there now vast numbers of television channels and radio stations, but there are also other (almost unlimited) sources of information and entertainment, many of which are associated with the Internet.

You do what?

Dividing people into groups according to their characteristics and behaviour is useful for studying how they behave at an aggregate level, so that one can design advertising campaigns and marketing strategies, but often one would like to know what will happen at the level of individuals. How will *this* customer react to this offer? What should we do when *that* borrower misses a repayment instalment? Which other products would *this* customer like? And so on.

To answer such questions, we must get down to the individual level and collect data and build models that tell us how individuals behave. This is done by studying how people have behaved in the past, and using this to construct models for how they, and others, might behave in the future. Thus, for example, we can collect information on the type of purchases people make in supermarkets, perhaps observing that people who buy sun-dried tomatoes also often buy bottled olives. We can then use that information to make predictions about the future purchasing behaviour of people who buy sun-dried tomatoes. We can also use it to decide what other new products the supermarket might stock: perhaps focaccia bread would also go down well with this group of purchasers.

Although, nominally, the scanning process at the supermarket checkout is to enable the total bill to be calculated, in fact, the details of all of the purchases are stored for later analysis. The scanning process involves reading a unique barcode on each item purchased. Barcodes are numerical and/or alphabetical codes that uniquely label objects. The most popular codes use black and white bars of differing widths. They are scanned by a laser reader, and then the code is looked up in a database to identify the object and its characteristics, including, for example, its price. Each barcode contains a start and stop character, so it does not matter which way it is read – the electronics in the reader can see which comes first and reverse the order of the characters if that is necessary. Some other forms of code, called matrix or stacked codes, use a two-dimensional array rather than the line of bars. They are not as widely used as barcodes at present, though they are gaining in popularity. There are several hundred different coding schemes in use, but the codes in most common use on retail products are the Universal Product Code (UPC) system in the US and the European Article Numbering (EAN) system in Europe and elsewhere.

Studying data on behaviour patterns also means one can forecast likely future trends. Are people taking more holidays overseas, as air ticket prices have fallen? How are events such as terrorist atrocities likely to affect international and even national travel? What about knock-on effects: reduced air travel means reduced need for aircraft catering and reduced need for taxi journeys to and from the airport. Observations on people's credit card usage patterns will enable one to predict how much they are likely to spend on Christmas presents. Indeed, close examination of the sorts of purchases they have made will enable highly targeted promotional strategies – it will reveal whether the purchaser has a family with small children, for example. One might paraphrase the old adage that 'We are what we eat', and update it to 'We are what we buy.' Very useful for business, and even for the government, juggling its economic strategy, but a perhaps a gross invasion of privacy?

A very simple example of studying behaviour patterns arose as part of a data mining exercise I carried out with Gordon Blunt, a former postgraduate student of mine. We were studying transactions made in petrol stations by people using credit cards. A simple plot of the sizes of these transactions for a sample of customers over the course of a year showed some unexpected behaviour. One of the diagrams from this study is reproduced in figure 4.1. The horizontal axis shows the size of the transaction, in pounds sterling. As can be seen, there are anomalously high counts at £5, £10, £15, £20, £25, £25, £30, and £40. It seems that some people, perhaps a surprisingly large number, choose to limit their purchase to a round number of pounds (even though they are paying by credit card). That this explanation was the correct one was supported by closer examinations of the values around the peaks. The peaks were very sharp to the left (the £4.99, £9.99 etc. were very much in line with the rest of the data), but there was a slight excess of values just above the peak: values such as £5.01, £10.01, etc. were slightly larger than one might have expected from the body of the data. It is as if people had tried to hit £5.00 and £10.00 but just occasionally overshot.

In fact, the behaviour of our petrol purchasers was even more subtle than the above suggests. The rounding to £5, £10, £15, etc. is clear from the diagram, but subjecting it to a statistical magnifying glass revealed the same phenomenon repeated at a smaller scale: people also tended to round to the nearest pound, so that, for example, £13.00 was more common than £13.04. Indeed, more surprises were to come in our investigation. There were also peaks, even smaller than those at whole numbers of pounds, at 50p: smaller, but just as striking when the data were displayed appropriately. In fact, we found such peaks, of diminishing size, at all levels in the data: smaller than the 50p peaks, there were peaks at 25p. Still smaller than these were peaks at 10p, 20p and so on. All this is very interesting, you may say, but surely such obsession with minutiae can have little practical consequence. Not true, as figure 4.1 also shows. There

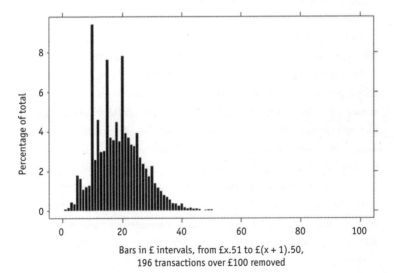

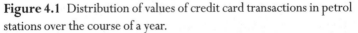

Figure 4.1 Distribution of values of credit card transactions in petrol stations over the course of a year.

are two other large peaks in this figure which I have not yet mentioned. These are at £12 and £18. They have arisen because an astute marketer and data analyst has already recognized the tendency people have to round their purchases to multiples of £5 or £10, and has induced people to spend more by offering them a free torch if they spent a little more. Some of our further explorations of this surprising data set are described in a paper which Gordon and I published in volume 12 of the *IMA Journal of Management Mathematics*.

The idea of studying data on behaviour does not solely apply to shopping, of course. The principle applies in all walks of life. Take, for example, the observation that three recent crimes in a small town have several very similar characteristics – that the perpetrator tends to behave in the same way. This suggests both that a serial offender is on the loose and how they operate, and might lead to indications of the sorts of situations in which they will strike and where they will strike next. This is not far-fetched: someone who has successfully robbed three banks is unlikely to switch to being a house-breaker, a mugger, or a car highjacker for his next job. After all, he seemed to know what he was doing with the bank robberies.

A controversial example of predicting likely future behavioural patterns is forecasting which children are most likely to commit crimes during their adolescence, and then graduate to a criminal lifestyle. Some fairly accurate predictive models can be built, based on data from past criminals and on data describing the infant and childhood circumstances of individuals. Tough ethical questions arise here: should society incarcerate such children when they reach the age of ten as protection against their likely criminal future? As I said before, it's all a question of balance: this time, of individual rights against society's well-being.

As a more detailed illustration of how deeply such behavioural modelling can go, let us look at credit scoring in the retail financial services industry – in deciding whether someone is a good risk for a mortgage, credit card, small loan, car finance, a store card, a current account, an overdraft, and so on.

Fifty years ago, if you wanted a loan – to buy a car or to extend your house, for example – you would have gone to your bank manager. You would have donned your best suit and tie, and he (for the manager would have been male in those days) would have used his personal knowledge of you to make a decision. Were you a respected member of the community? Were you reliable? Was your job secure? Did you have property which could be claimed to cover the cost of the loan if things went pear-shaped, etc. Gradually, however, banks became larger, and a separation grew between the manager and the customers. It was no longer feasible to rely on personal knowledge. People travelled more, and the number of requests for financial products began to explode. In part, this was handled by the introduction of revolving credit operations, such as credit cards, where individual requests for loans were met automatically, and where the key decision was about whether to grant an entire credit line on which you could draw at will and without further decision by the bank. But even these key decisions could no longer be made in the traditional way: in 2002, there were over 500 million credit cards in Europe, and personal interviewing of applicants would clearly have been impossible.

Furthermore, legislation gradually tightened. It became illegal to discriminate on the grounds of whether the applicant was male or female – that is, to include gender as a factor in the decision on whether a loan should be given. The only way to be sure that such excluded indicators are not used is to have a formula, a recipe, or an algorithm for deciding whether or not a loan should be granted which specifically excludes gender.

These various pressures meant that automatic systems, based on *scoring* applicants using data describing them, had to be used to drive the decision-making process. Points were awarded according to whether they owned or rented the house they lived in, how long had they been at that address, how old they were, how long they had been with their current employer, whether they had ever previously defaulted on a loan, and so on. Such systems were amongst the first developed, and are called *applications scoring* systems, because they assess applications for products.

Initially, people – both customers and bankers – were suspicious of such systems. How could a cold sum of numbers describing a person make accurate decisions about whether or not they were good risks? Comparative experiments were carried out to compare human and score performance. Their broad conclusion was that the objective statistical scoring methods had the edge: moreover, they also had significant practical advances. They did not tire, they did not make poor decisions because of a poor night's sleep, and they were not susceptible to personal prejudices irrelevant to creditworthiness. They did have the disadvantage that they were not effective in handling anomalous cases (e.g. the new CEO, who has just arrived from abroad so has no local credit track record, lives in temporary rented accommodation, and has obviously been a short time in their current employment, all of which are high risk factors), but that was a very minor issue on the overall scale of things – and in any case, such anomalous cases could be flagged for closer human examination.

Such objective methods also had one other advantage, one which was to prove decisive. An algorithm which is explicitly written down and applied to numerical data derived from the customers can be modified and experimented with. We can explore how the performance is changed if we double the weight applied to age in the model, if we mark rented accommodation as twice as risky as owner-occupied accommodation, if we slightly change the importance accorded to time with current employer, and so on. By such means we can incrementally improve the model, perhaps only a little at each step, but over time the little improvements add up, and the result is a significantly improved model. This sort of improvement just is not possible with subjective human approaches. They are too ill-defined and subject to extraneous influences. And the key to all this is data.

Nowadays, after several decades of research and development, such objective scoring systems are highly accurate. They are widely used in all sorts of decision processes, not merely for deciding whether or not someone is likely to default. For example, they are used for setting interest rates, deciding if someone is likely to take up an offer, if they will churn – that is be 'loyal' or switch to another supplier – and so on.

These ideas have also been significantly developed, beyond making one-off decisions, into *behavioural scoring*. Behavioural scoring monitors how people use their current account, credit card and so on over time. Behavioural scoring can be used for fraud detection, deciding whether to increase someone's credit limit, deciding whether to offer someone an additional financial product, and a host of other areas. Such models are based on the data arising from the series of transactions a customer makes: on what a customer does. When one remembers that (for example) credit card data includes what you bought, where you bought it, when you bought it, as well as the purchase price, it becomes very clear that personal financial transaction data are extremely informative about the actions and life of an individual.

Politics

It is probably unnecessary to remind anyone, especially anyone who has been living in Germany, the UK, or the US in the last few years, that elections are often close-run things. This means that a few votes here or a few votes there could tip the balance in a different direction. Since only one person can be elected to each office, the tiniest electoral majority translates into a complete dominance of the post. The fifty-three per cent of the Electoral College who supported Bush in the 2004 US Presidential election and the forty-seven per cent who supported Kerry translated into 100 per cent election of Bush as President. This means the system is exquisitely sensitive: a slight shift of proportions in the Electoral College would result in a 100 per cent shift in outcome of the election. In Germany's 2005 election, Angela Merkel won 35.2 per cent of the vote, whilst Gerhard Schröder won 34.3 per cent, but only one of them could be Chancellor. In the 2005 UK General Election, Labour won fifty-five per cent of the 646 Parliamentary seats, while the Conservatives won thirty per cent, but the respective percentages of the votes were thirty-five per cent and thirty-two per cent, much closer than the seats. Clearly the distribution of voters across seats is such as to give Labour an advantage. One could imagine an extreme situation in which party A had a majority vote, but this was almost entirely concentrated in one seat, so that party B won most of the seats.

When things are so delicately balanced, how the voters are distributed across polling regions is just as critical as persuading people to vote for you. What matters is getting your message to the right voters. Effort spent in a region where the vote is expected to be split 80:20 is wasted effort, but effort spent in a region where the vote is expected to be split 51:49 could make all the difference.

There's more to it than this, however. People are different: they may agree with your position on immigration, but disagree with your position on taxes. If you know this, if you have enough information on an individual voter, you can target your vote to match their interests. You can gloss over your tax plans and play up your immigration policy when canvassing. If you know that the crucial voters tend to watch a particular TV channel, then you can target your advertisements appropriately. (Apparently, in the 2004 US election, those most likely to vote for Bush preferred the Golf Channel, while those most likely to vote for Kerry preferred the Game Show Network.)

It used to be the case that what mattered was what the voters knew about the candidates, so they could decide who to vote for, but this has changed. What matters now is what the candidate knows about the voters. With this knowledge, the candidate can decide who to speak to and what to tell them. This has become known as 'microtargeting' or 'political sharpshooting'. As John Gertner (2004) put it in a *New York Times* article, *the very, very personal is the political*.

Personal data of this sort has become big business – after all, what could be bigger than major national elections? Companies have been set up specially to compile and collate such data: very much a case of Big Brother watching you.

Knowing all about you

Data on each and every one of us are stored in a wide variety of databases. In fact, let's make this more personal: data on *you, the reader of this book*, are stored in hundreds of databases. Governments collect and store individual data from censuses, electoral rolls, tax returns, company records, and so on. Banks and financial organizations know all about the things and services you buy, where your money comes from, whether you are reliable in making repayments, whether you max out your credit card, and so on. Market research organizations collect data on all sorts of behaviour, from reading and TV viewing patterns, to attitudes on animal rights and religious beliefs. Telecommunications companies know where you have travelled, via your cellphone records, when you were at home using your computer and what websites you have surfed. Health services know all about your medical history, treatments and likely future health. And it goes on.

It is probably obvious from your bank statements and credit card records that financial data are stored in great detail, but the same applies, and will apply even more in the future, to other collections of data. Take medical records as an example. The UK is creating a new National Health Service database containing all details of *everyone's* medical records: the *Care Record Service*. The spine of the system will contain name, address, NHS number, date of birth, and clinical information such as allergies, adverse drug reactions, details of accident and emergency visits and a link to a lower level of data. This lower level will contain more detailed information, stored locally, such as medical conditions, medication records, operations, test results, X-rays, scan results, etc. The aim is that health workers anywhere should be able to gain instant access, twenty-four hours a day, seven days a week. This will enable emergencies to be treated without wasting time, to prevent accidents happening when someone is given a medication to which they are allergic, a reduction in waiting time for X-rays to be delivered, and so on. Access to such information has to be restricted, of course, and will be by means of a Smartcard and PIN system. Prescriptions will be prepared and sent electronically via a messaging system, and patients will be able to access and update their own records. It goes without saying that the data will follow the patient, so it can be wherever it is needed.

In isolation, such databases give a narrow window on your behaviour and what sort of person you are: when it launches, the Care Record Service will tell all about your medical and health history, from cradle to grave. A whole new ball game begins when different databases are linked.

Record linkage is the process of matching records from different databases. So, for example, if we had access to the Care Record Service database and a supermarket transaction database, we could match your eating habits, via your supermarket shopping patterns, to your medical records, in a series of steps using your credit card number and the address it was registered at. The power for good this would bring is immense. It would bring a tremendous boon to epidemiologists, trying to discover those factors which predispose us to illness: it would have been simplicity itself to discover the dangers of smoking had such a system existed fifty years ago. More recently, an Australian study has used record linkage to link long-haul flight records with deep vein thrombosis records: patients admitted to hospital with venous thromboembolisms were matched to records from long haul flights to reveal that the annual risk of thromboembolism is increased by twelve per cent if one such flight is taken annually. In the US, the Centers for Disease Control is organizing the *National Violent Death Reporting System*, to link data about violent deaths from medical examiners, coroners, police, crime laboratories and death certificates, which will provide answers to questions about how, when, and where such deaths occur – with the obvious aim of identifying and alleviating the causes. Another British example is that the police can now use a database provided by the Association of British Insurers to instantly check whether a car carrying a particular registration plate is properly insured. With cameras automatically reading number plates, it is but a small step to have fines or court summonses sent out automatically.

Credit bureaux (credit reference agencies, credit registries) are amongst the most well-known commercial organizations which merge information from various sources. Banks and other financial bodies are naturally reticent about divulging data on their customers to the competition, but credit bureaux can act as a central clearing house, keeping records of whether customers have paid their bills and loan instalments on time. By acting as a central repository, they overcome the problem of a credit card issuer seeing only the customer's behaviour with that card. A more global view shows all the cards the customer has. This is beneficial both to the banks involved, and also to the customer, since he or she can be protected from over-committing themself.

Great opportunities for public good, perhaps, but do such systems also imply an invasion of individual privacy? Even worse, what happens when there is a mistake in the linkage? When the records from two different people are accidentally merged, for example. This happened to retired bus driver Frank Hughes, when another man with the same name was matched to him. Former workmates who had attended his funeral were later shocked to see him walking down the street. A shock for his friends and a surprise for Mr Hughes, but perhaps fairly minor on the global scale. Not so the warnings about record linkage from the TAPAC report, which says (Minow *et al*. 2004, pp. 37–38):

> One of the most significant of these issues concerns the
> significant difficulties of integrating data accurately. Business
> and government have long struggled with how to ensure that
> information about one person is correctly attributed to that
> individual and only to that individual ... According to the General
> Accounting Office, the government already suffers significant
> financial losses from its inability to integrate its own data
> accurately.

Accurate record linkage requires a unique identifier: something in the record describing a person that is unique to that person, and that is contained in both databases. Often such a unique identifier will consist of several fields taken together. Thus, an individual's name may not be a unique identifier (just type your name into Google and see what happens), and neither may an individual's address (there may be more than one person living at your address), but the two together may be unique. Even this is not always the case, however, and it might be necessary to add another field – date of birth, for example. Even then, mistakes occur. Identifiers which are supposed to refer to the same person may not do so, and different identifiers may refer to a single person: am I David J. Hand, David Hand, D.J.Hand, or D.Hand?

Record linkage work is probably most advanced in the US, where information on such things as property transfers, birth records and a wide variety of other sources have been cross-referenced and linked to provide an awesome level of detail, but other countries are catching up fast. In the UK, *The Times* reported on August 4 2005 that 'HBOS, Britain's biggest mortgage lender, is pressing the Government to force local authorities to provide banks with details of council tax arrears [in a drive to improve credit scoring predictions of who is a poor financial risk].' At a lower level, receipts for goods paid for by credit card often replace some figures by a series of X's. However, unless the same figures are always obliterated, a collection of receipts enables the credit card number to be discovered – be careful how you discard old credit card slips.

Sometimes a little thought allows data relevant to the overall good to be collected without compromising individual privacy. For example, the UK Government is considering replacing road and fuel tax by tolls, in which one pays for each mile travelled. The toll rates would range from 2p per mile on the quietest roads to £1.34 on the most congested. Pilot schemes are already beginning. This is all very well in principle, but to ensure that people pay the right amount, a satellite tracking system would be introduced to track each vehicle. From another perspective, this is simply a system for seeing where each driver goes, which represents a clear invasion of privacy, a concern which has been expressed by some motoring groups. To overcome this, an alternative system has been proposed, in which cars are fitted with cameras which could read the number plates of the vehicles in front. The number plate, and the details of where and when it was read, would be stored on an onboard computer, and transmitted to a roadside receiver, which would compare it with a database of who had paid to travel on what road. By this means, only those who had failed to pay would be identified.

If you think this sounds far-fetched, requiring a substantial technological advance, then think again. There is already a congestion charge system ringing the centre of London, which requires drivers to pay £8 every day they enter the area. The system is controlled by a ring of cameras which automatically read the number plates of all vehicles entering the zone, and compare these with a computer file of who has paid the charge. The only difference between the proposed toll system and the congestion charge system is that the cameras in the latter are fixed in place.

A delicate balance

It is clear that deciding what data will be stored about individuals, who will access it and how it will be used are all delicate issues. Balances have to be struck so that the maximum good and the minimum unnecessary invasion of privacy is achieved. Technology is advancing so fast that the law finds it difficult to keep up. Data residing on a server in a one country may be searched and analysed from another, so that it is not clear which legal system applies. The fact that we have both data *protection* laws (protecting not the data, but the individuals whom the data describes) and *freedom* of information laws demonstrates the attempt to strike a suitable balance. In particular areas, special restrictions apply, for example, the illegality of using gender as an indicator in making credit decisions in banking.

Different people have different perspectives on what is an ideal balance. At one extreme, little or no personal data would be stored, and a moratorium would be declared on the development of new ways of looking at such data. As we have seen, this would have terribly detrimental consequences: such data are a source of immense potential good for humanity – as individuals and in the aggregate. At the other extreme, no restrictions would be placed on the data and its analysis, but the scope this would give for invasion of privacy would be excessive.

Perhaps the bottom line is that more and more data about individuals are being collected anyway, whether one likes it or not, and we need to develop ways to cope with that: trying to hide our collective heads in the sand is not an option. The situation is just the same as with progress in any other scientific or technological area: you can't stop progress. Stem cell research, nuclear physics and genetic engineering provide other examples. If *you* don't do it, someone else will. All you can do is attempt to steer it in morally and ethically proper directions. Exactly the same applies to knives and hammers, which can be used for good or bad.

Jerrold Nadler made the same point when he appeared before the United State's Technology and Privacy Advisory Committee in November 2003: he said the 'question isn't whether technology will be developed, but rather whether it will be used wisely'.

In the context of data mining, the Executive Committee of the ACM Special Interest Group on Knowledge Discovery and Data Mining said, on 20 February 2003:

> Some of the people who are not knowledgeable about data mining
> technology are apparently equating data mining technology to
> guaranteed violation of civil liberties ... We believe that the debate
> about technology's potential for violating civil liberties should be
> focused solely on the misuse of stored data, not on anything else –
> not on data mining technology, database management technology,
> the Internet, or even the creation of a national database as
> envisioned by the TIA initiative.

CHAPTER FIVE

Modern Data Science

If the sums do not add up, the science is wrong. If there are no sums to be added up, no one can tell whether the science is right or wrong.

Donald Laming

What is modern data science?

We have seen that data are not merely numbers: numbers alone tell us nothing about the real world. For numbers to tell us something useful, for numbers to be data, we need something more. One vital thing we need is the metadata – the data about the data, which tell us what the data mean and set it in a context of relationships with other data items. But, even with metadata, the data are of limited value. For data to be useful, we have to convert it into information. That is, we have to *analyse* the data, with a view to using it to shed light on questions or topics in which we are interested. Put another way, we have to extract those aspects of the raw numbers that are useful for our purposes. In a scientific study, we are aiming to 'extract nature's secrets': to boil down the raw data to yield information about the causal mechanisms or relationships which led to them. To reveal why the stars behave as they do, how organisms came to be the way they are, why people act the way they do, and so on. In a commercial context, we aim to extract information from the data, which we can then use to run our corporation effectively, so that we do not overstretch our resources, but also so that we use them as efficiently as possible.

Modern data science has multiple aspects. Our ultimate aim is to analyse the data, but there are other necessary steps before we can do that. We must be able to store our data, and store it in a form so that we can access it efficiently and easily. If we have a large data set, describing billions of objects (think of an electronic astronomy catalogue or a database describing credit card transactions from a large bank), then without effective storage and access tools we are lost before we start. Once we have powerful tools for storing and accessing our data, then we can begin to think about manipulating it – about analysing it to extract something useful from it.

Because there are many different aspects to modern data science, numerous subdisciplines have developed, each concerned with some part of it. Two obvious examples of this are *database technology* and *statistics*. Database technology provides effective tools for organizing, storing, accessing, updating, and maintaining the consistency of data. Statistics provides tools for summarizing data and drawing conclusions about the process or mechanism which led to the data. Other relevant subdisciplines include *machine learning*, *data mining* and *pattern recognition*. Machine learning is concerned with constructing programs that analyse data to understand how the data arose so that appropriate future actions can be taken – to learn from experience and to generalize. Data mining analyses very large data sets, either to summarize them in convenient ways, or to find anomalies within them. Statistical pattern recognition aims to find relationships in data so that objects may be classified correctly. It will be obvious that there is considerable overlap between these various subdisciplines. In the past, this has led to some tension ('we, in our intellectual community, knew about that long ago – you are merely rediscovering the wheel'), but to a large extent these battles have died down as researchers have begun to appreciate that their broad aims are the same, but with different emphases.

One example of these differences in emphasis is the role played by *models* and *algorithms*. If one examines almost any modern statistics book or paper, one will see the word 'model' appearing. This word, and the concept it represents, has become ubiquitous in statistics. To a statistician, a model is a summary description of the data.

There are various different kinds of models. For example, a model may be based on some kind of underlying theory. That is, the form of the equations defining the model may be based on the way one believes the objects behave in real life. To take a particular case, one might believe that the pressure of gas in a closed container will depend on its temperature, and set up an equation in which pressure is proportional to temperature (letting an analysis of data decide the constant of proportionality). Or one might believe that people will take more driving risks if forced to wear seatbelts, and one might produce a model based on this theory, in which the accident rate is multiplied by some factor when seat belt legislation is introduced. Models of this kind are called *iconic* or *mechanistic*, since they are representations of an underlying mechanism.

In contrast, other models are simply descriptions of a set of data. We might have no idea why the risk of defaulting on a bank loan should be related to the colour of the car one drives, but if analysis of a set of data that includes information on loan defaults and on car colour shows that there is such a relationship, then it will be included in our model. Models of this kind are, perhaps not surprisingly, simply called *descriptive* models.

The term 'data model' is also used in a rather different sense by the database community. In statistics, a model is a description of the data as it occurs, summarizing its distribution, relationships between the observed characteristics, and so on, whereas in database technology, a model is a description of the structure imposed on the data. One of the founders of modern database theory, Edgar Frank Codd, defined a data model as having three components: a collection of data object types, a collection of integrity rules, which say what sort of objects can appear in the database, and a collection of operators that can be applied to the objects (for retrieval and other purposes).

If models are ubiquitous in statistics, in machine learning the central role is played by algorithms. An algorithm is a description of a process, or recipe, for doing something. In the case of data analysis, an algorithm will tell us how to process an input set of numbers to produce an output. The emphasis is on what you do with the numbers, rather than a description of them. If you look in machine learning textbooks, you will often see pseudocode descriptions (simplified versions of computer programs) of algorithms, telling you how to process the data to reach a conclusion. This is in contrast to the preferred mode of representation of the statistical community, whose models are represented by mathematical equations.

Having said all that, one should not lose sight of the fact that the model and algorithm approaches are two sides of the same coin. Both are needed in order to produce an effective analysis. All that differs is the perspective from which one looks at things.

Modern data science, then, is concerned with extracting useful meaning from the raw data – with peering through the fog of the numbers to discern useful structures. The sort of meaning that will be useful, the kinds of structures one will be looking for, will depend on the problem and the context. Because these are unlimited, so too are the methods of data analysis. We might want to build a tool that will enable us to predict the future: what will happen to the economy next year? Is a hurricane or earthquake likely in the next few months? We might want to build a tool for decision-making: should I prescribe drug A or drug B? What will happen if I change the school curriculum? We might simply want to understand what is going on: how are the clusters of galaxies related? Why do people turn to crime? At the root of all such questions lies the analysis of data. Part of the aim of this chapter is to take a lightning overview of modern tools for extracting meaning from data. However, data must be collected before one can analyse it. It is probably pretty obvious that the nature of the data determines how easily it can be used to answer questions, how accurately it can answer those questions, and indeed what questions it can be used to answer. Thus, the data should ideally be collected with the questions one wants to answer in mind. In particular, one will tune one's data collection procedures so that those questions can be answered as cheaply and as accurately as possible. The next section outlines data collection methods that ensure that these criteria are met.

Collecting data

To obtain accurate answers from the analysis of data, it is obvious that the data must be of high quality. Care must go into collecting the data. Take missing values, for example. At best, if some of the values are missing, then less accurate results are likely to be obtained; at worst, it means that the results can be positively misleading. Trying to find out the mean income of the people in a town from a survey that included only those available for home interviews on a Monday morning would almost certainly lead to an underestimate, because those being paid higher salaries would probably be busy at work earning those salaries, so their data would be missing. A classic example of this is a famous *Literary Digest* poll for the US Presidential Election of 1936. The *Literary Digest* had used polls to successfully predict the outcomes of the five previous Presidential elections. The magazine predicted that the 1936 Republican candidate, Alf Landon, would comfortably beat the Democrat Franklin D. Roosevelt in 1936. In fact, FDR won a landslide victory, gaining forty-six of the forty-eight states, with sixty-three per cent of the vote. What went wrong was that, although the sample of people polled by the magazine was very large, it was also biased. Over ten million questionnaires were mailed out, but they were mailed to people who had both automobile registrations and were listed in telephone books: this biased things toward the wealthy. Huge numbers of FDR supporters had neither, so that their opinions were missing from the data. The sampled population was distorted so that it did not properly represent the population of voters. Largely as a consequence of this miscalculation, the magazine soon folded.

The key point here is the (non)representative nature of the sample. In fact, if the sample is properly representative, then one can obtain accurate estimates with only a few thousand responses. One doesn't need anything like the millions which led to the disastrous miscalculation in the 1936 election. Data collection tools for ensuring that a properly representative sample is obtained are described in the section on survey sampling. Using these tools enables one to obtain estimates for the minimum cost, and, perhaps even more importantly, to state how accurate the estimates are. One can say how likely they are to be wrong.

In other situations other data collection strategies are necessary to ensure accurate estimates at minimum cost. For example, if we are studying the behaviour of people in stressful situations, the fragility of rocks, or the growth of crops, we can carry out an experiment in which some of the potential influences are kept fixed while others are varied (in these examples, potential influences might include, respectively, the nature and context of the psychological stress, the humidity and temperature of the rocks, the nature and amount of the fertilizers being used). There are complex and subtle ways of controlling for such influences, through which one can sometimes obtain quite extraordinary amounts of data. Tools for designing such experiments are described in the section on experimental design.

Sometimes special complications arise when collecting data. For example, the very presence of an observer taking measurements or asking questions can change the behaviour of the things being measured; this is especially the case when the things being measured are humans. Various subtle data collection strategies have been developed to cope with such difficulties. Concealed cameras are one obvious approach which can be used to study behaviour in some circumstances, but ethical considerations need to be taken into account. For sensitive questions, the method of 'randomized response' has been developed. It is most easily illustrated by an example. Suppose we want to ask 'Are you still a virgin?' but we suspect that the peer pressure of modern society will mean that many who are still virgins will lie. To get round this difficulty, we toss a coin. If the coin comes up heads, we hand the respondent a card with the question 'Are you still a virgin?' written on it. If the coin comes up tails, we hand them a card with the question 'Did you have toast for breakfast today?' on it. We do not know which card each receives. From another study, which just asks the 'toast for breakfast' question, we can estimate the proportion in the population who had toast. Then, from the total number answering yes overall (whichever question they happened to receive), it is easy to work out how many answered yes to the virgin question. Note that the key thing here is that we cannot say anything about an individual's response. We know they answered yes or no, but we do not know what their question was.

In many situations, electronics has revolutionized data collection. In particular, automatic measuring instruments surround us. Typically without our knowledge, they are silently collecting data on the air temperature and quality, on the flow of people through the metro system, on what you have bought and where you bought it, and so on. Such automatic systems have many advantages over direct human measurement. They are typically more accurate, they take measurements constantly, not just at those times when the human can get to the site, they can measure things in inhospitable environments or awkward locations, and they give the result immediately. Telemetry – the transmission of data over long distances, by phone, radio, etc. – means that the measurements can be taken anywhere. An extreme example, of course, is space probes measuring climatic conditions on other planets. It is largely the development of such automatic systems which has led to the increasing numbers of very large data sets which are now collected. In place of the human slowly measuring and recording each individual value, the electronic instruments simply, silently and anonymously, and at an incredible rate, map the values into a computer database.

Survey sampling

Most readers will have encountered surveys in some form or other. You might have been approached by someone with a clipboard while you were walking along a street (though sometimes these are sales operations disguised as surveys). Or perhaps you have been contacted by researchers conducting postal or telephone surveys. Surveys are a way of obtaining information about entire populations by examining just some of the members of those populations. The ideas have been most developed in the context of studying humans, but they apply generally to any situation in which we would like to obtain information about an entire population.

To illustrate, suppose that we want to estimate the average household income in a country. In principle, of course, we could simply ask every household in the country what its income is, and then average the figures. In practice, this could be very difficult and expensive: indeed, it might even be impossible, perhaps because the distribution of income in the population was changing even while the operation was being carried out. After all, carrying out such an operation would take a significant amount of time, and the people whose incomes we learned early on in the exercise might well have moved on, been promoted, or lost their jobs by the time we reached the end. Clearly the exercise would be much quicker and much more useful if we only needed to approach a small sample, rather than the entire population. It would also be much cheaper. If we could get an accurate answer from just a thousand people, instead of the entire population of ten million, then we have the potential to reduce the cost by a factor of ten thousand. This would mean we could carry out many more such similar exercises: our money would go much further.

This is the trick to survey sampling: instead of approaching the entire population, we will merely approach a sample of households and aim to extrapolate from them to the entire population. The key is to find some way to do this, with confidence that our extrapolated value gives an accurate value for the entire population. There are several fundamental ideas that make this possible.

One is that we have to be careful about which households we approach – which ones we include in our sample. Obviously it would not be a good idea to include only those living in the most expensive cities in the country, or only those living in poor rural areas or inner-city slums. We really need to draw a *representative* sample. We could try to do this by deliberately choosing the households we included in our sample: by *purposive sampling*. This is all very well, but it has risks. It is entirely possible that we will overlook aspects of the households we have chosen that in some way make them nonrepresentative of the entire population of households. For example, even if we chose some households from richer areas and some from poorer areas, perhaps we subconsciously selected households that had two parents, and underrepresented single-parent households. If this sort of thing happened and we were unaware of it, there would not be much we could do about it.

Although this is true, there is a brilliant way in which the problem can be sidestepped. Instead of deliberately choosing which households to include, we *randomly* choose them.

At first, this probably seems a totally bizarre and absurd idea. If we cannot control our sample so that it is representative by *deliberately* choosing which households to include, how on earth can *randomly* selecting them help? After all, if we randomly select which households to include, surely there is a chance that we will obtain an unrepresentative sample. There is even an outside chance that all of the households included in our sample will come from the rich area. All of this is true, but – and this is the secret – if we randomly select our sample, *we know the probability that these extreme selections will arise*. In fact, we can calculate the chance of any configuration of households being chosen. From this, we can work out how likely it is that the sample we have chosen is extreme and nonrepresentative. The brilliance of the idea is that it makes chance – randomness – work in our favour. It is a sort of data judo, making the opponent's strength and weight work to our advantage.

How does being able to calculate the chance of the different selections help us? Here is a simple example. Suppose that our overall population consists of six Americans and six Europeans, and that we wish to choose a sample of size six. If we randomly choose our sample in such a way that all possible samples of size six are equally likely, then the chance of getting a sample consisting of Americans only is approximately one in a thousand (in fact, it is exactly one in 924). Likewise, the chance of drawing a sample consisting of Europeans only is about one in a thousand. However, the chance of drawing a sample which has three Americans and three Europeans is approximately 433 in a thousand – nearly a half (in fact, it is exactly 400 in 924). We are much more likely to draw a balanced sample than a grossly unbalanced one. This huge difference is because, while there is only one single sample consisting of six Americans (there are only six in the population), there are many different samples which consist of three Americans and three Europeans.

We can extend these ideas to work out the chance of drawing a sample that had five Americans and one European, and so on. We can thus say how likely (or perhaps I should say how unlikely) it is that we will observe a sample which deviates substantially from the true proportion in the population, and exactly the same sort of calculations apply to other characteristics of the people in the population.

This example involved very small numbers. With realistically large numbers the chance of a substantial deviation from the true population value becomes microscopically small. What is interesting, however, is that, if the population is large, this probability depends on the size of the *sample*, not on the size of the population. The probability is essentially the same with a sample size of 1,000 whether the population is a million, ten million, or a hundred million. This means that we can get accurate estimates for huge populations based only on relatively small (and hence, cheap and quick to collect) samples.

Of course, this basic idea of drawing a sample requires some elaboration to make it effective. First, in order to be able to draw samples with equal probability – as each possible sample of size six in the above example was equally likely to be drawn – we need a list of everyone in the population. Such a list is known as a *sampling frame*. Such things are not always available, and while techniques have been developed to sidestep the problem, in general the better the sampling frame, the more accurate the results. Certainly, a poor sampling frame can lead to serious errors. The reader will probably have already have thought of these dangers in the context of magazine, radio, or television 'surveys' which invite readers, listeners, or viewers to send in their opinions. Clearly those most concerned with the issues are most likely to respond. Imagine a magazine 'survey' which asked the question 'Do you reply to magazine surveys?' How many 'no' answers do you think it would get? Would the magazine's editor then be correct in concluding that almost everyone replied to such surveys, on the grounds that everyone who responded had said 'yes'?

Although the idea of random sampling means that we can say how confident we are that our results do not deviate substantially from the true population value, there is always the chance that they might.

While, by taking a large enough sample, we can make this chance as small as we like, there are also other ways we can reduce this chance. One way is to *stratify* the sample. In stratified sampling, one splits the population into relatively homogeneous subgroups, according to what one thinks the value of the characteristic one is exploring will be. So, for example, suppose we want to estimate the average income in a town. We could stratify it into regions according to how expensive the houses were. Then we would take a separate random sample in each of these regions, combining them to yield an overall average. This prevents the possibility that our overall sample might consist solely of people from the rich areas or of people from the poor areas.

There are many other variants on simple random sampling. For example, if one is going to send out interviewers to talk to the people selected for the sample, one would like to minimize travel costs. It would then be an advantage to select people who live near each other. Likewise, in a survey of schoolchildren, it might be more convenient to sample every child in selected classes, rather than children from every class. Methods of *cluster sampling* have been developed to cope with this. Stratified sampling and cluster sampling can be combined: we might stratify into church schools and non-church schools, and then use a cluster sample design in each stratum. There are also ways of taking advantage of any other information one already has about potential respondents.

All of these tools, and others, lead to more efficient, cheaper, quicker and more accurate ways of collecting data about a population.

Experimental design

In Chapter 3, we explored the nature and origins of science, and saw how it is based on the notion of evidence – that is, of data – as the driving force for understanding the world about us. Key to that development is the idea of the experiment: the *deliberate* manipulation of the things being studied so that one can obtain accurate information about them. Naturally we want to gain the maximum information we can for the smallest effort (which may mean lowest cost, or quickest time, or least number of objects experimented on, etc.). To achieve this, we need to structure our data collection, our deliberate manipulation, very carefully. The idea of attempting to structure our data collection has produced some remarkably elegant and powerful ideas, leading to great advances in science, agriculture, manufacturing, and commerce. The area goes under the name of *experimental design*.

There are several notions central to experimental design. One is that of *control*. If we wish to know whether action A causes effect B, we must compare the outcome when action A is taken and the outcome when it is not taken. For example, in a *clinical trial*, the kind of experiment conducted to investigate the effectiveness of medicines, we must compare people who take drug A with a *control group* of people who differ in no other way but for the fact that they do not take the drug. Only then can we be sure that any effect is due to the drug and not to some other influence. This elementary fact is often overlooked in nonscientific statements about causal relationships. Comparing a group of young people who took the drug with a group of older people who did not would not enable us to separate the influence of the drug from that of age. Comparing a group of people who took drugs A and B with a group who took neither A nor B would not enable us to separate the effect of drug A from that of drug B. Any effect we observe could have been caused by either drug.

An instructive example of the dangers of not including a control group arose with an operation involving a porto-systemic vascular shunt in hepatic cirrhosis. This operation was introduced in 1945. Some twenty years later an investigation of sixty-five studies of its effectiveness showed considerable support for the procedure in those studies which did not include a control group, and very limited support for it in those studies which did include a control group. In fact, the operation was no better than the alternatives. The trouble is, of course, that once a medical procedure has become established it is very difficult to question it and to suggest that some patients might be treated by an alternative procedure.

The control group in such a trial might receive a standard treatment, so that we can determine if a proposed new treatment is more effective, or they might receive a 'placebo' treatment, which is known to be inactive, so that we can determine if the new treatment has any effect. Of course, it is important that the patients do not know which treatment they are receiving: if they did know, there would be a difference between the treatment groups in addition to the treatment. It is entirely possible, for example, that patients who knew they were receiving a new drug might act differently given this knowledge. Indeed, such effects are well established. Studies in which the patients do not know which treatment they are receiving are called *blind* studies. More generally, and equally important, it is usual that the doctor administering the treatment does not know which drug is which, or she might treat the patients differently (either consciously or subconsciously). This is a *double-blind* study.

You may have noticed that I referred to the *group* of patients receiving the drugs above. Different people react differently, and even the same person reacts differently at different times and under different conditions, so that there will be range of responses to a drug. This means it would be unwise to test the drug on just one person. Instead, a group of patients receive the drug and we look at the average response. Exactly the same sort of principle applies in other experimental contexts: if we are measuring the pliability of plastic as it is heated, we want to allow for measurement error and slight differences in constitution, so we will take several measurements and average them: if we are testing the efficacy of a fertilizer, we will apply it to several plots of land and measure the plant growth; and so on. These repeated measurements of the same phenomenon are termed *replicates*. As we shall see below, part of the magic of data analysis is that, by taking enough replicates, we can make the results of our analysis as accurate as we wish.

Of course, we must be very careful in how we *select* the patients in a trial such as that above, and also in how we *allocate* them to the two groups. As far as selection is concerned, if we intend our results to apply to all ages, we must ensure that all ages are represented in our study. It would obviously be a mistake merely to test teenage males, if we hope also to draw conclusions about the effect of the medicine on female octogenarians. This is the sample selection problem discussed above. Usually a process of *controlled random sampling* is used to select the patients. 'Controlled' here means that there are typically some exclusion criteria and also often some restrictions to ensure that the population of interest is covered. For example, we might exclude pregnant women to avoid any risk to the unborn foetus, and arrange that roughly equal numbers of males and females are included so that we cover both sexes. This is the notion of stratification described above.

Once we have selected a sample of patients from the population of interest, we must decide which of them will receive the placebo and which the active treatment. Once again, the notion of randomness comes in: we randomly allocate the patients to each of the two treatment arms, control and active. Again there are two reasons for the random aspect; firstly, that valid statistical conclusions can be drawn, as described below, and secondly, that it eliminates sources of conscious or subconscious bias. For example, if a researcher deliberately chose how to assign the patients to the groups, then it is possible that he or she might choose to assign patients which differed in some systematic way to the two groups. Perhaps the more seriously ill patients might be assigned to the active treatment, and the less seriously ill to the inactive control. In such a case, we would say that disease severity was *confounded* with treatment: we would be unable to disentangle the two possible causes of any difference between the two groups that we did discover. An illustration of this is described in Chapter 6, in the famous Lanarkshire milk study.

There is no doubt that controlled experiments have revolutionized the way we live. They have advanced our understanding from Francis Bacon's warning about 'dreams of our own imagination' to a proper grasp of how nature and the universe functions. The application in a medical context, the randomized clinical trial, has been described as the greatest advance in medical research ever.

Of course, the design described above is just about the simplest experimental design there can be. It just involves two groups, a baseline or control group, with which the outcome of the active or treatment group is compared, but the ideas have been taken much further, leading to immensely powerful tools for extracting nature's secrets.

An obvious extension to the simple two group study is to use several groups of patients, perhaps to compare different doses of a drug to see which level is the most effective. A more sophisticated extension is to compare several *factors* simultaneously. In the above example, the drug being tested represents a factor, and, in the simple two group study it occurred at two *levels* – zero, and whatever dose was used in the treatment group. However, we might be interested in studying several factors

To illustrate with something entirely different, suppose we are running a plastics injection-moulding company and making plastic household goods, and that each time we make a product we can grade it on a numerical quality scale. Our machines have various controls, including those for the pressure at which the plastic is injected, the temperature at which it is injected, the time over which it is allowed to cool, and so on. Each of these represent factors in the process of making the goods, and each may change the quality of the final product. For simplicity we shall suppose that each of these factors has just two levels, low and high, and we would like to discover, for each factor, which of its two levels produces the superior product.

We could use the same sort of design as in the clinical trial to study each of the factors separately. So, keeping temperature and time fixed, we could vary the pressure, carrying out a two-group study and producing a set of products at low pressure and another set at high pressure. Keeping temperature and time fixed means that any differences we observe must be attributable to the one thing that differs between the two sets of measurements – the pressure. Note that since, just as with people, we should expect some variability in the result, it would not be sufficient to produce just a single product at each of the levels of pressure. Finally, we could compare the average quality scores of the two sets of products to see which pressure yielded the superior output. Suppose, for this illustration, that we made four products at each level of pressure. Then we would have to make eight products in all.

Similar designs, also requiring eight products, could be run to estimate the effects of temperature and time.

Thus, at a cost of making a total of $8 \times 3 = 24$ products, we can tell which of the levels of each factor is better.

This is all very well, but there is a much cleverer way to obtain the same information, a way that yields just as accurate results, but in much less time and for much less expense. With the three factors of pressure, temperature and time, each at two levels, there are eight possible combinations of level. These are pressure low, temperature low and time low; pressure low, temperature low and time high; pressure low, temperature high and time low; and so on, ending up with all three factors at their high levels. Examination shows that amongst these eight combinations, four of them have pressure low and four have pressure high. Moreover, these eight runs are balanced: for the product made at pressure low, temperature low, time low, there is a matching product at pressure high, temperature low, time low. These have the same levels for temperature and time, so only pressure (and inaccuracy of measurement) can account for the difference in quality of the products made by these two runs. Now this is exactly what we need for the two group comparison described above. From these eight measurements, we can estimate the effect of pressure on the quality of the product, with four meas-urements in each group.

So far, so good: this is much the same as the simple two group experiment in which we looked solely at the effect of pressure. However, if we look more closely at the eight treatment combinations we see that, amongst them, four also have temperature low and four have temperature high and these are also balanced: for the pressure low, temperature low, time low combination, there is a matching pressure low, temperature high, time low combination. Now the only difference in quality must arise from the temperature difference. So exactly the same applies as in the case of pressure: we have a simple two group experiment, with four measurements taken at each level of temperature. We can estimate the effect of temperature on the quality of the product, with four measurements in each group.

Finally, of course, as you will have guessed, the same applies to time. The eight measurements include four taken at the high level of time and four at the low level, so that we can test the effect of time on the quality of the output, again with four in each group.

To summarize this, in place of the twenty-four products and measurements we would have to produce if we did things in the obvious way, judicious balancing of the different levels of the treatments when we make the products means that we can obtain exactly the same information from just eight products and measurements: a third of the time and a third of the cost.

Designs of this kind, which involve multiple factors varied simultaneously in a carefully controlled manner, are called *factorial designs*.

In fact, we can go further than this. It is possible that the effect of pressure (for example) will depend on the level of temperature. So, for example, perhaps high pressure produces better products than low pressure when the temperature is high, but high pressure produces worse products than low pressure when the temperature is low. This sort of relationship between pressure and temperature is called an *interaction*. The separate experiments on each of the factors do not allow us to test for such an interaction, but the artful combination of levels of the treatments in the three factor factorial design described above does allow us to test for the presence of interactions.

Extraordinarily enough, it is possible to go even further than this: it is possible to reduce the number of combinations at which one must make a product to even less than eight in a *fractional factorial* design, and yet still test each of the three factors To the uninitiated, this appears to be magic: it seems as if one is obtaining information for nothing. But you can't get something for nothing: the conservation laws of physics tell us this, as does the 'no free lunch' law of economics. The truth is that, with three factors there are eight possible treatment combinations, and we can only reduce the number of products we make from eight while still obtaining information about each of the factors if we make some assumptions about what is going on. In particular, we have to assume that there are no interactions: that the effect of each factor is independent of the level of the other factors. If we can do this (perhaps we believe it to be the case because of earlier experiments), then we can save even more money and time.

There are many other sophisticated, and indeed subtle and clever, designs for obtaining data as accurately and efficiently as possible. *Latin squares* allow two factors which have more than two levels to be tested without looking at all possible combinations. *Greco-Latin squares* extend this to three factors. Matched pairs studies reduce the effect of inaccuracies of measurement by applying different treatments to the same individuals (not always possible, of course: if the treatment is the way in which a child is taught to read, one can hardly apply different treatments to the same child). *Cross-over designs*, common in medicine, are an example of this sort. *Repeated measures designs* extend these ideas to multiple treatments. These and many other kinds of design have been developed to meet the needs of data collection in the modern world.

Storing data

The modern avalanche of data manifests itself in two ways. One is the rate at which we are deluged by data: numbers are thrown at us from all directions, at rates far too great for us to absorb. Automatic data acquisition technologies and electronic measuring instruments produce numbers at vast rates, but these numbers don't simply vanish into the ether. They are stored, and this is the second way in which the modern data avalanche manifests itself. There are more and more larger and larger databases being accumulated, describing the world about us, and also describing us.

These numbers are not simply thrown into a metaphorical bin when they arrive, jumbled up higgledy-piggledy at random in a disorganized mess. Rather, they are organised into some kind of structure – so that they can be recovered, searched for, and generally subjected to analyses of various kinds that will reveal things about the data and their source. The organizing structure forms a *database*, which is looked after by a piece of software called a *database management system* (DBMS). Such a program performs multiple functions. It allows you to access the piece of data you want – and with large databases, speed is vital. It ensures that new additions to the database are added in a consistent way. It arranges that adjustments or corrections to the database do not lead to contradictions, so that errors are not introduced, controls who has access to the data, and at what level (all of the data, or only parts of it; summaries or the individual data items?) and ensures that, if a program crashes while in the midst of analysing the data, no data are lost, and so on. The DBMS is a housekeeper for the data.

Large databases cannot be stored in the main memory of a computer. Instead, they have to be stored in secondary memory devices, such as hard discs. This means that access times are not always as fast as one might like. Moreover, accessing different pieces of data will depend on where in the database they are stored. Thus, designing the database structure with its ultimate use in mind is important: pieces of data that will be accessed frequently should be quickly locatable. Moreover, if different users are intending to use the data in different ways, it may be beneficial to store copies of the data, structured in different ways. Again, it is important to ensure that they are consistent, and do not contradict each other.

Data may be organized in various ways. Different organizational structures are best suited to different problem domains. Three important structures are *hierarchical*, *network* and *relational* structures.

Hierarchical structures take the form of trees. Entities at each level are subdivided into components. Thus, for example, in a database describing the employees of a corporation, we might have a record for each employee, and this might be split into a component describing their personal characteristics (age, salary, address, etc.)

and a component describing their position and role in the corporation (department, job, etc.). Each of these components may then be
further subdivided. Perhaps the department component is split into
subcomponents describing the section (containing their number in
the section, when they joined the section, etc.), their duties in the
section and so on, in a progressively more detailed hierarchy of data.

Hierarchical structures are very quick and convenient for some
purposes. It is easy, for example, to locate information about a particular individual. However, some tasks are more difficult. For
example, obtaining overall descriptions (e.g. the age distribution of
everyone in a particular section) is a time-consuming operation. Of
course there are ways round this, but only at the cost of increasing
the complexity of the control structure, and sacrificing some of its
elegance. Essentially, these solutions lead to a structure in which
items of data are connected by more than the simple links of a tree: a
network or graph structure is used. Now, instead of two items of
data being connected by a single path (which, in a tree, would be to
go from one data item towards the root of the tree until a common
ancestor was identified, and then to work out towards the other item
of data), there may be multiple paths between data items. Although
a network structure permits searches of a wider variety to be conducted more rapidly, it is at the cost of an altogether more complex
system. This means that a more complex management system is
required in order to maintain its integrity (to prevent contradictions
from creeping in) and also that more storage is required (to store the
wider variety of links).

Finally, we come to relational models. Relational databases have
become very important, and most databases nowadays adopt this
approach. The origins of such databases lie in the mathematics of set
theory – and this solid theoretical basis means that proofs (of things
such as consistency) are easier than in other structures. Instead of the
notion of hierarchy, with some groups of data items being contained
within other items, the main idea is that of *relationship*. A relationship
can be thought of as a table, in which the rows represent entities and the
columns properties of the entities. Different tables present different

'views' of the data, describing entities at different levels, and different properties of those entities. For example, one view may contain departments as the entities, while another may contain individual employees. Because this fundamental concept is so simple – and so familiar – the basic principles of relational databases are easy to learn. Of course, there is more to it than these few sentences imply. In particular, one needs to find convenient tables in which to store the data: a single large table, in our example containing employees as rows and all the different items of information about them as columns, may permit anything to be pulled from the database, but it would be horribly inefficient and unnecessarily large (e.g. descriptions of each department would be replicated for each employee in a department). To put things into context, perhaps it is worth commenting that databases containing thousands of tables are not unusual.

Statistics

At last, having stored our data in a form which will permit us to access and manipulate it, we come to data analysis itself. The oldest data analytic technology is statistics, having origins dating at least as far back as the seventeenth century, but being properly developed as a discipline in its own right throughout the twentieth century. With the appearance of the computer, other data analytic disciplines such as machine learning, pattern recognition and data mining began to appear. This section describes statistics, and the next, the newer data analytic technologies.

The development of statistics occurred in three phases. The first is that which took place prior to about 1890. The Manchester Statistical Society was established in 1833, the Royal Statistical Society in 1834, and the American Statistical Association in 1839 – clearly the 1830s were an exciting time for statistics. However, the early discipline was very different from the modern discipline. In particular, it was not until almost 1900 that mathematics began to appear (and a glance in any modern statistics journal will reveal that it is now an intensely mathematical discipline). In 1885, a remarkable paper by Francis

Ysidro Edgeworth introduced a wide variety of modern mathematical statistical concepts, including significance tests, use of the median as a measure of average, parametric versus nonparametric tests, and the tendency of the distribution of a mean to approach a normal distribution. This paper marks something of a watershed. Soon after it appeared, three papers were published in 1893 that introduced the mathematical statistical concepts of the bivariate normal distribution, the correlation coefficient, standard deviation and skewness, and from then on it was all downhill (or uphill, depending on how you look at it), with more and more mathematics appearing.

The second phase is characterized by this mathematization of the discipline, and took place between about 1890 and 1960. It yielded an immense flowering of the mathematical underpinning of modern statistics, with the development of a tremendous range of ideas and methods of stunning beauty and elegance.

Finally, the third phase is marked by what has happened since about 1960. Over this time, by virtue of the impact of the computer, the discipline of statistics has metamorphosed yet further. Whereas in the first half of the twentieth century it was probably legitimate to regard it as a mathematical discipline (or, at least, mathematically based), things have now changed. Nowadays, it might be more accurate to regard statistics as a computational discipline. Certainly modern statistics would not be possible without the computer.

The first objective in a statistical data analysis is to summarize a set of data. One might have various aims in mind when producing this summary. For example, one might simply want to gain an understanding of a data set – an understanding that would be quite impossible by looking at the numbers individually. Alternatively, one might want to find a formula that would enable one to predict likely future observations. Or perhaps one wants to calculate the probability that an extreme value would be observed in the future (in insurance, for example). There is an unlimited set of possible reasons why one might wish to summarize the data. Each new problem will come with its own reasons, and even the simplest of data sets can be summarized in multiple ways. The way one chooses must depend on what one wants the summary for, or what question one is seeking to answer. This is true even for as simple a measure as the 'average'.

An average can be calculated in different ways. A common measure is the 'arithmetic mean', or just '*mean*' for short, calculated by adding up the data values and dividing by how many of these values there are. Another common measure is the *median*, the value such that half the data have smaller values and half have larger values, and there are other measures as well. Which is appropriate depends on what one is trying to do. Suppose my data set consists of the results of a thousand measurements, with 500 of them taking a value of 10001, 499 of them taking a value of 9999, and just a single measurement taking a value of 0. The arithmetic mean has the value of

$$(10001 \times 500 + 9999 \times 499 + 0 \times 1) \div 1000 = 9990.001$$

This means that all the measurements except for the 0 value are larger than the 'average'. Some people would find this surprising, but perhaps that just means that the arithmetic mean is not the right average for the problem. The median value for these data is halfway between 9999 and 10001, equal to 10000. Whether this the right average to use will depend on what aspect of the data one is trying to convey – on why one is summarizing it.

Of course, a competent data analyst would have spotted the unusual *outlier* of 0 and investigated it to see if it was an error. After all, it is very different from all the other measures. Perhaps something went wrong with the instrument when that measurement was taken, and perhaps that value should not be included in the calculation. In case you think this is a contrived situation, with little bearing on reality, Chapter 7 gives a similar example concerning the salaries of baseball players prior to the strike in 1994, showing how the right choice of average, again between arithmetic mean and median, can be crucial: in that case, millions of dollars hinged on it.

In these examples, the discrepancy between the mean and the median arises because the data are *skewed*. The differences between the values at one end of the distribution are much greater than the differences between the values at the other end: in the numerical example, the difference between the two smallest values (0 and 9999) is much greater than the difference between the two largest values (both equal to 10001). Such situations are quite common when there is a natural lower bound on the values (e.g. baseball players cannot earn less than $0 a year, house prices cannot be less than $0, etc.) – although then it is the larger values that are more spread out. But questions about the adequacy of an average as a summary of a set of data can arise in even more straightforward situations. For example, if a dataset has values ranging from one centimetre to a million centimetres, one might be more interested in this range than in a single summary 'average value'. Telling someone that the mean is 10,000 cm, for example, is perhaps more obscuring than revealing. Likewise, reporting that one's average temperature is fine when one hand is on fire and the other is frozen solid is rather missing the point. These examples suggest that perhaps more than one number might be used to summarize a data set. Perhaps an average and a measure of *dispersion* – of how widely spread the values are – might helpfully be used in conjunction.

Just as with averages, so there are various measures of dispersion. Familiar measures are the range (simply the difference between the largest and smallest values) and the variance (the arithmetic mean of the squared differences between the individual data points and their mean). The standard deviation is the square root of the variance, useful because it will be measured in the same units as the raw data, rather than in squared units (after all, what does a squared kilogram mean?).

In fact, one can continue this, summarizing a data set in terms of multiple summary statistics, taken together, and each describing a different aspect of the data. But, of course, there comes a limit: if one has too many such statistics, it rather defeats the object of producing a simple summary of the data.

So far, all the discussion has been about summarizing a single set of values. Often, however, in both commerce and science, the aim of data analysis is to relate different variables. Do the data show that income goes up with number of years of education? If we decrease the lengths of coffee breaks, will the output of a firm go up? Is the mass of a star related to its luminosity? And so on.

Regression analysis is a fundamental tool used for describing the relationship between a 'response' variable and a 'predictor' variable. Given data consisting of many pairs of observed values of the predictor and corresponding response, regression analysis approximates the response values as a simple multiple of the predictor plus a constant term. For example, we could predict a child's future adult height from its birth weight using *height* $= a \times weight + b$. Here a is the multiplying factor and b is the constant, where both values are estimated from the adult heights and birth weights of a group of people.

In more elaborate forms, regression does something similar when there are multiple predictor variables and a single response variable: for example, if we want to predict the output of a factory from its number of employees, average employee wage and number of items it produces each week. This opens up all sorts of additional possibilities for shedding light on the data and how it arose. For example, we can answer the questions of which predictor variable is the most important for predicting the response.

Regression is an asymmetric sort of concept: some variable is singled out as a response, and we are interested in how its values differ for different values of the predictor or predictors. The related concept of *correlation* does not single out one variable as a response, but simply describes the extent to which two variables tend to take large values together or small values together. Human height and weight are correlated, for example: taller people tend to be heavier than shorter people. One can see from this example that the relationship is not perfect – some tall people do not weigh very much, whereas some short people weigh a lot. The relationship is a statistical one, describing the general pattern of how the two variables tend to go together.

In some situations, the available data may consist of all possible observations in a situation, and one might want to understand their characteristics. For example, one might want to know the mean age of employees in a company, or how many men there are over fifty working in the sales force, when one has data on all the employees. In other situations, however, the aim is to make an *inference* from the values of just a sample of data to some population from which the sample has been drawn. This population may be real (e.g., all the people who live in a particular town; all the purchases people made from your company last year), but in other cases it will be notional (e.g. all the possible measurements I could make of the weight of a rock, which would not all be identical because of slight errors in the measurement procedure; or the answers 'all men' would give to a particular psychological measurement procedure). In both cases, we imagine that there is a real population, with a real mean, a real variance, and so on, and our aim is to *estimate* those population descriptors using our sample of measurements. In general, population descriptors are called *parameters*.

Of course, our estimates will not be perfect – indeed, we should expect them to vary between different samples. I can estimate the mean height of men in the population by taking the mean of a sample of the heights of a hundred men, but surely we would not expect the result to be exactly the same for a different sample of a hundred. One of the arts of statistics is knowing how to choose good ways of estimating – good 'estimators' – and the ability to determine how far the sample-based estimate is likely to be from the true population value. In the case of using a sample mean to estimate a population mean, if the sample measurements are drawn randomly from the population, then it is possible to give precise numerical values for the probability that the sample mean will lie any given distance from the population mean. That is, we can calculate the dispersion of the distribution of the possible values of the estimate. One measure of how good is an estimator is given by the size of this dispersion. Clearly an estimator which yields estimates that are tightly bunched about the true value will be better than one that yields values that have a high probability of being very different from the true value.

We saw earlier that the variance – the mean of the squared differences between the data values and their mean – is a popular measure of dispersion. In fact, there is a close relationship between the variance of the values in a population and the variance of the means of samples drawn from that population. Suppose that we take a sample of size n. That is, we randomly take n objects from the population, measure their values and calculate their mean. Now suppose we repeat this many times, taking different random samples, all of size n, and so producing a whole distribution of sample means, with this distribution itself having a spread, a dispersion. Then if the variance of the values in the population is v, the variance of the distribution of means will be v/n. In fact, more than this, the mean of this distribution of means is the same as the mean of the original population.

These are exciting results: the first result means that we can make the variance of the distribution of mean values as small as we like, simply by choosing a large enough sample (making n large enough). Combined with the second result, it implies that we can estimate the mean of the original population *as accurately as we like*, simply by taking a large enough sample. The larger the sample we use, that is the larger the n, the higher the probability that our sample mean will be close to our population mean. Actually, if the population is finite (e.g., the number of people in a corporation) this result may be obvious – make the sample large enough and you are including almost everyone in the population. However, it also applies if the population is effectively infinite: the possible measurements I could take of the weight of a rock, for example. In particular, it means that, provided my measuring instrument is not 'biased' (as discussed in the next paragraph) even if it is very inaccurate, so that the results differ dramatically each time I take a measurement, I can still get as accurate a result as I want, simply by taking many measurements and averaging them. We are beginning to see the power of statistics as a tool for uncovering structure in the universe.

The arithmetic mean has nice properties – such as the inverse relationship between the sample size used in calculating the mean and the variance of the mean – but not all estimators have such attractive properties. In particular, sometimes estimators are *biased*. All this means is that the mean of such estimates is not exactly equal to the parameter one is trying to estimate. That is, on average there will be a discrepancy between the estimated value and the true value. Of course, if this discrepancy is small, it will not matter.

The idea that we are using the mean value calculated from a sample as an *estimate* of the mean value of a population generalizes. In regression, for example, we can imagine that the values of the multiplier *a* and the constant *b* calculated from the sample are estimates of corresponding ('true') values for the population from which the sample was drawn. But how do we calculate the sample values – how do we estimate the values of *a* and *b* from our pairs (of height and weight values in the example)?

One very important method is the method of *maximum likelihood*. This is a general method, which can be applied to any estimation problem. If one believes that the data arose from a population of values which has a distribution of a particular form then one can calculate the probability of getting the observed data. In general, the population distribution will have a form that varies according to the values of certain parameters (describing, perhaps, its mean value, its dispersion, its skewness, and so on). Then one can choose those parameter values that have the highest probability of producing the observed data. It would, after all, be a perverse kind of estimation procedure that resulted in parameter values that were very unlikely to produce the observed data!

This general principle can be applied to all sorts of situations, no matter how complicated, provided one is prepared to make assumptions about the population distributions from which the data arises. Often these assumptions follow naturally from the context, and it is not always crucial that they be very accurate, so the method is immensely powerful.

Estimation procedures tell us the likely values of parameters describing a model or theory underlying our observed data. Often, however, we simply want to answer more straightforward questions. Does training method A yield different performance scores from training method B on average? Is *this* type of aircraft cheaper to run than *that* type? Is there any relationship between taking vitamin supplements and longevity? And so on. Questions of this type are answered by statistical testing.

The basic principle behind statistical tests seems rather subtle when encountered for the first time. As an example, suppose we want to know whether two populations have the same mean on some variable (the performance score question above, for example). That is, we want to know if the mean score we would get if we trained everyone using method A is the same as the mean score we would get if we trained everyone using method B. We can also express this as saying that we want to know whether the overall difference in mean scores from the two methods, for the entire population, is zero. The data we will use to examine this will be the scores from a sample trained by each method. Of course, the means of *samples* of scores from people trained by each method are likely to differ, even if the means we would get if we trained the entire population by each method do not differ. That is, even if, over all the people we could possibly train, there was no difference on average, it is likely that the means of two samples (of size ten each, for example) would differ; after all, means vary from sample to sample. Now, suppose we assume that the overall populations have the same mean. Then (with a few additional assumptions about the shapes of the distributions) we can calculate how often we would expect to get a difference between the sample means as large or larger than the difference we actually observed. If our calculations showed that obtaining such a large difference between the sample mean was very unlikely, it would mean that either a very unlikely event had occurred (the population means were equal, and we just happened to draw samples with very different means) or that the population means were different (and a much more likely event had occurred).

This is the core of statistical testing. We adopt some assumption (in technical statistical terms this assumption is called the *null hypothesis*), we collect our sample data, we calculate some summary from the sample data, and we see how often such a value of the summary would be obtained if the null hypothesis assumption were true. If such a value would occur only very rarely, then either a very rare event has occurred, or our assumption is wrong.

In practice, of course, we attach a number to what we mean by 'very rarely'. For example, we might decide that an event that has a probability of only 1 in 100 is sufficiently rare to arouse one's suspicions.

This approach to testing whether a proposed model or parameter value is correct can be turned on its head to produce *confidence intervals*. When we discussed estimation above, we concentrated on finding a single good value or 'best estimate'; for example, the best estimate of a population average, the best estimates of the multiplier and constant in regression. But we also saw that these were only esti-mates – in general, since estimates are based on samples of data, they themselves will have a distribution, varying from sample to sample, and there is a chance that they could be far from the true value in the underlying population. If the estimator had nice properties, the chance that the estimates were substantially wrong would be small, but it would be nice to quantify that in some way. This is what confi-dence intervals do. Instead of giving single a numerical estimate for a parameter, they give a range of values. Now just as single estimates would vary from sample to sample, so the ranges would vary from sample to sample. Some would contain the true parameter value, and some would not. The width of the confidence interval is chosen so that the proportion of such ranges that contain the true value takes a specific value. For example, we might choose a 99 per cent confidence interval, so that 99 per cent of such ranges would contain the true value. This would give us confidence that the particular one calculated from our data contained the true value.

So far we have taken the view that the parameters of populations are fixed but unknown. All of the random variation was regarded as lying in the process that produced the sample: different samples would differ, and the role of the statistical methods was to capture and control that variation. The statistical tools described above are then regarded as 'good' if they had a high probability of yielding a result near to the true value (estimation), of rejecting a false null hypothesis (testing), or of spanning the true value (confidence intervals).

An alternative perspective is to regard the sample as fixed (after all, the data are the one thing we actually *do* know) and the unknown parameter as having a probability distribution. Having estimated this probability distribution of the unknown true parameter, we could say, for example, what we thought was the most likely value of the parameter, how likely we thought it was that the parameter lay outside some specified range, and so on. This alternative perspective on inference is called the *Bayesian* approach. The approach that is based on regarding the parameters as fixed (if unknown) and the data as varying from sample to sample is the *frequentist* approach. The latter half of the twentieth century saw a major controversy about which method was superior, with the statistical scientific journals containing some sharp and extensive exchanges. To a large extent this has now died down. The consensus (though there are still occasional die-hard protagonists on each side who see their approach as the one true method) seems to be that different approaches are suitable for different problems. The Bayesian school, in particular, has received a tremendous boost from the development of powerful computers. Without that boost, the ideas would have been of philosophical interest but of limited practical value, as they required some very tough calculations to be made. The computer has made Bayesian ideas fly.

The statistical ideas outlined above are basic ideas, but statistics has travelled a long way in the last few decades. Here are just a few examples of modern statistical data analytic tools.

A useful division of statistical techniques is into those aimed at exploration and those aimed at confirmation. An exploratory tool is one that simply seeks interesting structures in data sets. One has to say roughly what one means by an 'interesting structure' but the search is fairly unconstrained. In contrast, confirmatory techniques examine data sets to see if they contain particular kinds of structures – perhaps to test a theory which says that they will do so. Often particular statistical tools can be used in both exploratory and confirmatory ways, so that the distinction is not always clear-cut.

Cluster analysis is an example of an exploratory tool. The aim is to see if a population falls into naturally distinct groups, and to characterize those groups. Given a sample of objects from the population, we can calculate the dissimilarities or distances between those objects in terms of the data describing them. Then cluster analysis finds a partition of the sample so that similar objects lie in the same group, and dissimilar objects lie in different groups.

The ideas of cluster analysis are used in a wide range of different applications. For example:

- in medicine, one might want to know if an apparent disease is really two or more distinct types of disease. In the 1970s, for example, cluster analysis was used to tease apart the nature of depression, helping to distinguish unipolar depression from bipolar depression.
- in marketing and customer relationship management, cluster analysis is used to identify different types of customers, perhaps on the basis of their behaviour patterns. An example is the FRuitS system mentioned earlier, which used data from the Financial Research Survey to group people according to financial status, financial product holdings and personal financial behaviour.
- a classic example of the application of cluster analysis is in taxonomy of plants and animals. The overt features of the objects – insects of a certain kind, say – are measured, and then used to see if they fall into distinct kinds. So, for example, we might measure caudal width, antenna length and so on of ants, to see if the sample includes examples from more than one species.
- a more recent application of cluster analysis is in genomics and bioinformatics. For example, the tools are used to try to identify those genes that behave in similar ways, hoping that this will shed light on biological function.

A completely different statistical tool which is often used in an exploratory way is *principal components analysis*. When the data consist of measures on multiple attributes taken on a set of objects, one will generally find that the dispersions of these different variables differ – the ranges and variances of some might be much greater than those of others, for example. Moreover, typically the variables will be correlated. To take an example, an entomologist studying aphids (greenfly) might measure body length, body width, forewing length, number of spiracles, the length of the type I antenna, the femur length, number of ovipositor spines and so on. These variables will have different degrees of dispersion, with some varying dramatically between different insects and others hardly at all, and they are all likely to be related to a greater or lesser extent. Our aim is to summarize these data in a convenient way. Principal components analysis does this by finding the weighted sum of the measured variables, which itself has maximum variation.

To understand this, and to see why it is useful, suppose first that we simply calculated the mean of the variables for each insect. Then each insect would be summarized by a single number, and we could compare insects in terms of those single numbers. Larger insects, for example, would tend to have large body lengths, body widths, etc., and would therefore tend to have a larger summarizing mean. The mean would serve as a measure of 'size' in some sense. In calculating the mean for each aphid, we would simply add up the values for that aphid and divide by however many variables there were. It is as if we calculated a weighted sum of the variables, but gave them all equal weights (equal to the reciprocal of the number of variables). An alternative would be to give the variables different weights. In particular, we could weight the variables so that the weighted sum had the largest dispersion amongst all such weighted sums. (Obviously some kind of normalization is necessary, or one could make the dispersion of the weighted sum as large as one liked, simply by choosing large enough weights for all of the variables, but this is easy to arrange.) This weighted sum that has the largest dispersion is characterizing some key aspect of the difference between the aphids: it is the feature of the aphids in which they most differ, and thus a useful characterizing feature of the aphids.

In fact, this idea can be taken further. Given the weighted sum that maximizes the dispersion between aphids, one can then seek the weighted sum which maximizes dispersion *but that is also uncorrelated with the first weighted sum*. This is identifying some other key aspect of the difference between the aphids, an aspect quite unlinked with the first aspect.

In this explanation, we started by identifying the weighted sum on which the aphids differed to the greatest extent, but one could also start at the opposite end. That is, one could find that weighted sum on which the aphids differed to the least extent. Now this might not seem like a very useful thing to do in the context of aphids, but consider a different situation. Imagine a physics experiment in which a variety of measurements had been taken – mass, pressure, temperature, light intensity, etc. – for a set of objects. A weighted sum of these variables on which the objects barely differed is equivalent to saying that this function of the variables (this weighted sum) is a constant for the objects. Put another way, it gives us a relationship between the variables that is satisfied for all such objects. That is, simple data analysis using principal component analysis will allow us to discover the equations of physics directly from the data.

Another technique, which is related to principal components analysis, is linear discriminant analysis. Principal components analysis seeks that weighted sum of the variables which summarizes the data on a single set of objects in a convenient way. Discriminant analysis also seeks a weighted sum of the variables, but it seeks that weighted sum which maximally discriminates between the members of two groups of objects, in the sense that members of one group will tend to have large values of the weighted sum and members of the other group will tend to have small values. Given any set of weights, we can calculate the weighted sum of the values of the variables measured for each object – for simplicity, call these weighted sums the scores of the objects. If we have two groups of objects, we can then calculate the means of the scores in each group. An obvious measure of separability between the two groups would then be the difference between the mean scores: the larger the difference, the more separate are the groups. However, if the variances of the scores in the two groups are large, there might still be substantial overlap between the scores in the two groups: the group with the larger mean may include many small scores if the dispersion of this group is large. To allow for this, the difference between the means is divided by the average dispersion in the two groups. The set of weights which maximizes this measure maximally separates the sets of scores.

This sort of idea has two main applications. One is to study the weights themselves: a larger weight will mean that a variable is more important in distinguishing between the groups. A small weight, one which hardly contributes to the weighted sum at all, implies that there is not much separation between the groups on the corresponding variable. Thus, one can identify the key ways in which the groups differ.

The second way is for classification. Given a new object, and its values for the variables, we can now apply the weights to calculate its score and then decide whether it is closer to the mean of one group than the other. We can then classify it as belonging to the group with the nearest mean. This sort of idea is very widely used, in both science and commerce.

Much data involves repeated measurements of the same thing at successive points in time. We are all very familiar with this from the daily news reports of the Dow Industrial Index, the FTSE-100 index, and the NASDAQ index of share prices. These are examples of *time series*. Thousands of other economic series are updated every day (of inflation, GDP, GNP, unemployment, immigration to the USA, etc.), and the same ideas occur everywhere. Time series of temperature show global warming. Time series of IQ scores show a gradual upward trend (called the Flynn effect). Time series of birth rates show cyclical changes. And so on. Anything which is repeatedly measured over time produces a time series, and the study of how things change over time can be immensely enlightening – both for understanding what happened in the past, and for forecasting the future.

There are many methods of analysing time series, taking into account deep aspects of how the data are thought to have arisen. One may postulate a gradual trend over time, a seasonal change, other cyclical patterns, or correlations between successive measurements for example. Furthermore, different approaches to analysing time series data have been developed within different sciences. Signal processing in physics and engineering has led to approaches distinct from those in economics. The discovery of *wavelets* has led to dramatic progress in some aspects of time series data analysis.

Just a few statistical tools have been mentioned above. There are many, many more, all aimed at analysing data to read the message within: at condensing the details of the many separate numerical results to give answers to the questions one is interested in. Some of these tools are simple, but are so powerful they are used thousands, perhaps millions of times every day around the world (the *t-test*, regression, analysis of variance, scaling methods, graphical methods and many others are all examples), while others are extremely complex and can only be sensibly applied by real experts. Part of the reason for this huge breadth is simply that statistics is the oldest data analytic discipline. Its origins, and indeed most of its development, predate the computer. This fact leads to an interesting speculation: if statistics was invented now, given that we have such powerful computing machines at hand, would it be any different?

I think it would. I think that many of the tools we now use, which were originally developed at least partly to overcome computational difficulties, might not exist. If statistics were to be invented from scratch now, for example, I suspect that so-called *computationally intensive* methods (which sometimes have exotic names, such as the *jackknife*, *bootstrap* and *Markov chain Monte Carlo*) would be used immediately, and that older, perhaps mathematically more elegant methods – for calculating the variance of complicated estimators, for example – would not have been developed.

Machine learning, data mining, pattern recognition, neural networks and more

If statistics is the oldest data analytic discipline, several others have grown up, motivated, I suspect, not primarily by perceived inadequacies of statistics, but simply because the originators often did not know what statistics could do. The result is a set of disciplines which overlap considerably with statistics, but which all have their own unique emphases and ideas. The aim of this section is to look briefly at some of these.

One of them is *machine learning*, one definition of which is that it is 'the art of constructing programs which learn from experience'. Experience, of course, is represented by data, and learning from these data requires analysing and summarizing them. Small wonder, then, that statistics and machine learning have substantial overlap.

Many machine learning tasks can be conceptualized as inferring general descriptions from samples of objects. This, after all, is what we do when we learn about the world: we experience a few times that Y happens when we do X, and infer that Y will always happen when we do X (e.g., a baby learning that its mother will feed it if it cries). Sometimes, of course, we can be wrong.

A common subtype of such problems is when the sample consists of two classes of objects, those of a type that we are interested in and those that are not. We then really have a classification problem, in which the aim is to learn to classify objects as belonging or not belonging to the class of interest. If the descriptions are couched in numerical terms – as measurements on a set of variables describing the objects – then the situation is really one of linear discriminant analysis, described above, and this method can certainly be used to tackle such problems. There are also many other methods which can be used, some of which have been developed by statisticians, by machine learning researchers, by data miners, by experts in computational learning theory and by workers in pattern recognition.

Because there are so many such methods, and because the problems they tackle are so ubiquitous, I will briefly describe some of them.

We saw in the previous section that linear discriminant analysis attempts to find the weighted sum which maximally separates the classes. An alternative is to model the *probability* that an object with any given set of descriptor variables will belong to class 1 (with the complementary probability that it will belong to class 0). One drawback of a weighted sum of the raw predictor variables of the kind used in discriminant analysis is that it is not constrained to lie between 0 and 1, but (for large enough values of these predictors) could give arbitrarily large positive or negative values, so we cannot use a weighted sum for this purpose. One way to overcome this is to calculate a weighted sum and then transform the result to force it to lie between 0 and 1. A transformation which is often used for this is the *logistic transformation*, and hence the name *logistic discriminant analysis* for this method.

Both linear and logistic discriminant analysis have weighted sums at their core, and this turns out to be a major weakness. In particular, it means that they cannot tackle the *exclusive-or* problem. Suppose we have two predictor variables – call them a and b, each of which can take just two values, 0 and 1. This means that each object must take one of only four possible pairs of values, $(a,b) = (0,0), (0,1), (1,0),$ or $(1,1)$. Now suppose that class 0 objects take values $(0,0)$ or $(1,1)$, while class 1 objects take values $(0,1)$ or $(1,0)$. Then there is no way that a weighted sum can be constructed such that objects in class 1 all have higher scores than objects in class 0. It just cannot be done (by all means try it).

We can overcome this problem by taking nonlinear transformations of the variables before calculating the weighted sum. (The transformations have to be nonlinear, because linear transformations yield results that are equivalent to weighted sums.) In fact, the idea is often taken further and several weighted sums of nonlinear transformations of weighted sums of nonlinear transformations of weighted sums of … are used. The result is a score for each possible object and, if the weights are properly chosen, it can be arranged that all class 1 objects score higher than all class 0 objects. If it sounds complicated, it is. What I have just described is an elementary *neural network*.

Of course, estimating the parameters (the weights) in such models is a huge task and although these ideas were proposed in the 1960s, it was not until the 1980s (computers again) that such methods became useful in practice. Since then they have been explored in great depth, both theoretically and practically, and have developed into a very powerful tool.

One of the interesting oddities of the way science develops is that it often bifurcates, with multiple strands emerging from a single source. This is true in this area. Neural networks are one extension of the idea of weighted sums, but a completely different extension has also grown up. This is the idea of *support vector machines*. Again we start with a simple weighted sum of the raw predictor variables, which is not very flexible and cannot solve certain problems. So now we generalize it by including nonlinear transformations of the raw variables, moreover we include *many* such transformations – squares, logarithms, products etc. In terms of the descriptors given by these new variables, if we have enough of them, we will be certain to be able to separate the data by a weighted sum (unless two objects have *exactly* the same predictor values). The trouble is that the calculations will be very time-consuming. This problem is overcome by some very clever mathematics, which sidesteps the need to make the transformations explicitly. The result is that a much easier calculation is done which is equivalent to computing a weighted sum from the transformed variables.

Once again, computers are needed, but once again a very powerful tool has resulted.

Neural networks and support vector machines are relatively new inventions, but there are older methods which can be just as effective if used carefully. I will describe just two: *tree classifiers* and *nearest neighbour methods*.

One can think of classification rules as partitioning the possible set of objects according to the values taken by their predictor variables. Objects with predictor variables taking some combination of values are assigned to one class, and the remaining objects to the other class. Tree classifiers construct such a partition recursively, by looking at the individual variables one at a time. Suppose, for example, we wanted to assign people who had suffered a recent head injury into one of two classes: whether their future disability would be mild or severe. Suppose the predictor variables included age, score on the Glasgow Coma Scale (a measure of responsiveness to stimulation) and change in neurological functioning over the first 24 hours. A tree classifier could take the form:

If Glasgow Coma Scale score is greater than r
    classify as class *mild disability*
If Glasgow Coma Scale score is less than r
    If change in neurological functioning is greater than t
        classify as class *mild disability*
    If change in neurological functioning is less than t
        classify as class *severe disability*

where r and t are numerical thresholds. A new patient is first tested on the Glasgow Coma Scale. If their score is larger than r then they are predicted as having mild disability. If their score is less than r they are tested on their neurological functioning. If this test gives a result greater than t they are predicted as having mild disability, but otherwise as having severe disability. Of course, this can be continued with multiple tests, and a large tree, consisting of multiple decision points, can result. The result of all this is that future objects progressively work their way down the tree, from decision point to decision point, until they end up at one labelled with a class.

So, using such a tree is simplicity itself. Classifications can be made very quickly, and small trees (with few decision points) are easy to interpret. They provide a natural protocol for use outside a computer, and are widely used in, for example, medical decision-making.

But how do we construct such a tree in the first place? To find a tree, one takes data from a sample of patients with known predictor variables and for whom the true outcome class is known. Then one simply looks at all possible splits on all possible variables (in our example, all possible values of r and t), to find the variable and the split for which the division best separates the two classes. In an ideal world, for example, this split would have all the patients who really belong to the mild disability class on one side of the split, and all the other patients on the other side. Of course, we do not live in an ideal world, and typically there are patients of both types on both sides, though a greater proportion of the mild disability on one side, and a greater proportion of the severe disability on the other. We therefore repeat the exercise, looking at each side of this first split separately, trying to split the reduced set of patients into two groups that are better separated. And so we go on, each time splitting one of the reduced sets of patients. Eventually, we end up with regions (such as the region: Glasgow Coma scale less than r, neurological functioning less than t) which contain only (or largely) sample patients from one group (in this case, severe disability).

Computers are ideal for rapid searching and comparison, so although this sounds like a lot of work (perhaps there are hundreds, or even thousands of predictor variables, on each of which we want to consider 100 possible split points) such trees can be constructed very rapidly.

The nearest neighbour method is very different, in that it makes no attempt to summarize the initial data sample. To classify a new point one simply looks at those sample objects which are most similar to the new object, in terms of the predictor variables, and assigns the new point to the same class as the majority of these similar objects. For example, if 9 out of 10 of the objects most similar (i.e. 'nearest') to the new object all belong to class 0, one would assign the new object to class 0. Although this is clearly a very simple and elegant approach, it does have some computational burden – one has to search over all of the known objects to find those that are most similar to the new object, but again this is normally not an issue with modern machinery. Of course, one has to decide quite what one means by 'most similar' – similarity can be measured in many ways – but research has produced some highly effective measures. Methods of this sort are described by various names, including nearest neighbour, case-based reasoning and lazy learning.

I have described just a few of the methods used for constructing classification rules. Many others exist – as well as many variants of those described above. Different rules can also be combined in various ways, to yield even more powerful combinations. For example, we might combine two rules that tend to misclassify different objects, so that the overall combination is better. A more fundamental extension is to *average* over a large set of rules. For example, we could take a subset of objects from the sample with known classes and build a tree classifier on those. Then we could take another subset and build another tree classifier. We could repeat this 100 times. A new point would then be predicted on the basis of the average (or majority) prediction from these 100 tree classifiers. A version of this has been extensively developed by Leo Breiman of Stanford University. He calls it the *random forests* model and it seems to be very powerful. Such averaging, to yield *ensemble* classifiers, can be justified on the basis of Bayesian arguments, where the true model is taken to be unknown and we are trying to average over the distribution of possible models.

The methods described above all also have a clearly statistical flavour, and scientific papers describing extensions and generalizations of them can be found in the statistical as well as the machine learning literature. Much of the flavour of machine learning is inherited from the fact that data are stored in computers in databases, and much of these data are categorical in form. That is, the variables or characteristics describing the objects each take one of a limited number of categories. For example, in a database describing customers we might have 'type of accommodation' categorized as 'home owner', 'rental', or 'parental home', and we might have 'marital status' categorized as 'married', 'single', or 'divorced'. Often, even variables that are in principle continuous, such as age, are grouped into categories. A class of methods which reflects this categorical flavour is that based on *rules*. A 'rule' is a logical structure of the form 'If conditions A are satisfied then B follows.' 'If someone has a high income, and has previously bought products of this kind, then they are worth including in a mailshot.'

Rules can be found by analysing databases. In fact, *association analysis*, a cornerstone of modern data mining, is aimed at doing just that – at identifying configurations of object characteristics which typically imply that other characteristics will also hold.

Rules can also be strung together to deduce things about the data. If we know that 'If A then B' and also that 'If B and C then D', then if we find an object for which A and C are true, we know that D will also hold. Of course, typically in real applications things are more complicated than this, because we generally don't find things such as 'If A then B' but rather 'If A then B ninety per cent of the time'. Further complications arise from the need to ensure that rules do not contradict each other ('If A then B' and 'If A then not B') and that contradictions cannot be deduced within a set of rules. Furthermore, it is desirable to keep the rule system as simple as possible. The second of the two rules 'If A then B' and 'If A and C then B' is superfluous. This is, of course, a trivial example, and some deep mathematics and clever computer algorithms have been developed to try to ensure this in realistically complicated cases.

Rule-based systems lie at the heart of *expert systems*. They are also very closely linked to tree methods. A tree, after all, consists of a combination of structures of forms such as 'If a is greater than r, and if b is greater than t, assign the object to class 0'.

Data mining is a recently developed data analysis technology, also very much a child of the computer. Again it overlaps very substantially with statistics and machine learning, but again it has a somewhat different flavour and emphasis. In particular, data mining is especially concerned with very large data sets which were originally collected for some other purpose, and which are now being examined in detail in a search for other useful information. For example, we might study a database of supermarket transactions to see if there are patterns that would be useful for future marketing initiatives. Or we might study a database describing astronomical objects, to see if we could detect any anomalies.

Data mining is usefully divided into two types of exercise. The first is concerned with summarizing the databases, or parts of them, much as in statistical modelling. One difference, however, is that because the data were typically collected for some other purpose (the supermarket transaction data were collected so that the customer could be billed, not primarily so that the transaction patterns could be studied) they may have many problems: there may be missing data, distorted data, or other problems of the kind discussed in Chapter 6. A second difference arises from the sheer size of the data sets. In particular, many statistical models are fitted by iterative estimation procedures, and while these may be fine with a few thousand data points, they might be completely impractical with billions. One possible way round this is to base the model on a sample, and use methods of statistical inference.

The second type of data mining exercise is aimed at detecting anomalies in large datasets. Sometimes this will require a simple search for anomalies whose structure we know, while at other times it will require one to define what one means by an anomaly at the same time as searching the database. Examples of the first kind are the search through credit card transactions (probably in real time, as the transactions are made) to detect simultaneous use of a card in two separated geographical locations, or the search for several purchases of small electrical items on the same card in rapid succession. Both of these patterns are characteristic of fraud. This does not mean that if they happen then fraud has necessarily occurred, but they should raise suspicions. (Small electrical goods can easily be sold on the black market.) Other examples include the detection of a black hole when a star appears to move against the background stars, as its light is bent round the black hole, and the detection of the pattern characteristic of a petit mal epileptic fit in an electro-encephalogram trace.

Situations in which we have to define the anomalies as well as detect them are more difficult. A relatively straightforward kind, and one which has been explored in the statistical literature, is the detection of outliers. Outliers are objects that are very different from the others. Outliers are relatively easy to detect if they differ from the mass of data on a single variable (the example of the

0 amongst the 1000 values discussed on p. 133, for example) but it is entirely possible that they are not at all unusual on individual variables taken one at a time, but only on a combination of variables. A ten-year-old female human is not at all odd, and neither is a female human with five children. But a ten-year-old female human with five children is distinctly odd. Tougher still are anomalies which do not involve individual objects at all, but groups of objects. A single case of leukaemia is not odd, but five cases clustered in a small area should arouse suspicions. In situations like this, one postulates a background model and seeks departures from it. The trouble is that random variation will always produce some departures, and if one has a large enough data sets, such peculiarities are almost certain to occur somewhere. This is exactly the principle which underlies the so-called 'bible code'. This is a report of words and sentences being found concealed in the letters of the bible (if, for example, you take every fifth letter, or follow a more complicated sequence – the letters occurring after alternate gaps of five and two letters, for example). The trouble is that if one examines enough such sequences, in a text as long as the bible, and with no limitations on what concealed message is searched for, then one is almost certain to find something. The discovery of these 'hidden messages' is merely a consequence of large data sets and the rules of basic probability.

With more and more larger and larger data sets being collected, data mining is becoming increasingly important. It is used in analysing stock market movements, in clickstream analysis of web searches, in customer relationship management, in analysing text data, in bioinformatics and genomics, and in a huge variety of other areas. Effective search algorithms have been developed, but there still remain important questions. Anomalous patterns in the data are not that interesting if they turn out to be due to chance or to poor data quality. Neither are they interesting if, as so often turns out to be the case, they were obvious or well-known beforehand.

And even more

This chapter has been primarily concerned with how to extract information from data, but there are other, also very important, aspects of modern data science. One of the most important is how to communicate data: that is, how to transmit and receive it. Often the transmission channels are not perfect, so that there is a possibility that the data may become distorted in its journey. Tools have been developed to tackle this, by building in redundancy to the message. Moreover, often one wants to be sure that only the intended recipients can understand a message – so we need to encode it. Coding methods – cryptography – have become of vital importance, not merely in classical military and espionage applications, but also in commercial situations. This does not apply just to big corporations – it also applies to transmissions between computers and in your Internet purchases and Internet banking transactions. Unless people can be confident that their details are safe, they will not use these facilities. Significant advances have been made in recent years. For example, *public key* cryptography schemes involve two 'keys', the first of which is used for encoding a message – this key can be public – the second of which is used for decoding a message – this key must be private. The trick is constructing a system in which it is very difficult to deduce the private key from the public key, and methods for doing this are based on tough mathematical problems (such as factorizing very large numbers).

Communication is closely linked to ideas of data compression. If redundancies can be removed from data then it is quicker and cheaper to transmit the data. This is particularly important for data such as images, which involve large data sets. Television pictures, and pictures from space shots are compressed in this way (as I write this, we have just received pictures from Saturn's moon Titan). Perhaps a more familiar example is the Zip facility on your computer, which allows files to be stored in less space than the original.

Redundancies can be removed in various ways. If an image contains a large block of the same colour and intensity, it might be better to describe a single pixel and the shape of the area covered by such pixels. If the data consist of a string of symbols (such as alphabetical characters) then a code in which the most common symbols are encoded as short messages and the less common ones as long messages can be highly effective. For example, suppose that we have a long sequence built up of just the three letters a, b, and c in some order (for example, *cbbabbcaaaabababc* ...), and we know that on average about ninety per cent of the letters will be a and about five per cent will be each of b and c. Suppose also that we want to use binary encodings of these letters to transmit them (that is, our system is like Morse code – it has only two distinct states). Then one possibility would be to encode a as 00, b as 1, and c as 01. These would be uniquely decoded at the receiving end. However, the average number of binary digits per letter in a message would be $2 \times 0.9 + 1 \times 0.05 + 2 \times 0.05 = 1.95$ bits. If, instead, we coded a as 0, b as 10, and c as 11, then the average number of bits per letter in a message would be $1 \times 0.9 + 2 \times 0.05 + 2 \times 0.05 = 1.1$ bits, resulting in a very substantial saving.

Displaying data is another a key aspect of modern data science. You will probably be familiar with the aphorism that 'a picture is worth a thousand words'. This is because the human eye has evolved to understand images. However, a picture is worth a thousand words only if it is well constructed. Poor graphics can conceal the truth, or even disguise it.

Graphical methods can serve several roles. A key one is that illustrated in the above saying – for communication, but another is to help with understanding the data. Consider figure 5.1 below, produced by Dr Mark Kelly, one of my former graduate students, in a study of bank loan data. Here, the jagged curve shows a plot of the proportion of new loan customers each week who repaid their loans by direct debit, and it is immediately obvious that there was a change in behaviour halfway through the four-and-a-half-year period. This would not be so clear from the raw data.

Figure 5.2, produced by Dr Gordon Blunt in a data mining study of credit card transactions, shows the cumulative amounts spent in

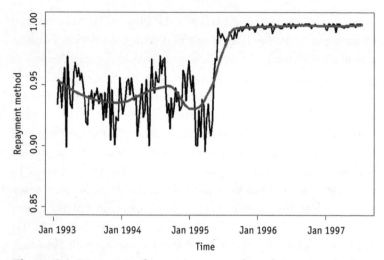

Figure 5.1 Proportion of new customers each week repaying a bank loan by direct debit, over a four year period.

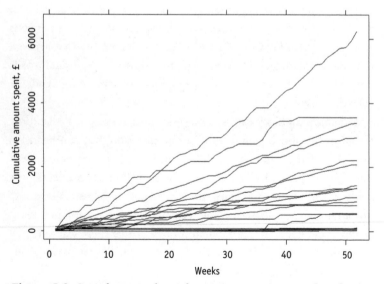

Figure 5.2 Cumulative credit card transactions in supermarkets for several different customers.

supermarkets on the card over a period of a year for a number of different customers. Most of the curves show a slightly irregular but persistent rise over time, but a handful show an apparent sudden cessation of use. This pattern is immediately obvious from the graph, but it would be very difficult to detect merely by looking at a table of numbers.

These examples are typical of the way graphical tools were used until the late twentieth century: but then the computer revolution occurred, and its impact on graphical methods was just as great as its impact on other aspects of data analysis. In particular, colour displays on screens arrived. Even more exciting, dynamic and interactive graphics became possible. Now one can watch one's data evolving before one's eyes. High-dimensional data can be probed inter-actively, plotted this way and that, effectively instantaneously, in an effort to tease out its secrets.

CHAPTER SIX

In Data we Trust

No good model ever accounted for all the facts, since some data was bound to be misleading if not plain wrong.

James Dewey Watson

Introduction

There is an old and familiar acronym: GIGO. It's short for *Garbage In, Garbage Out*, and signifies that if one's original data are nonsense, then the conclusions drawn from analysing those data will also be nonsense. This was just as true in ancient times as it is now. A farmer would soon have starved if he had consistently overestimated the number of sheep in his flock, and a modern space mission will fail if the wrong units are used. In fact, this is exactly what happened in 1999, with the Climate Orbiter Mars probe, when Lockheed Martin Corporation used pressure data based on pounds while the Jet Propulsion Laboratory used newtons. The result of this confusion was that the probe failed, probably because it entered the Martian atmosphere too low and burned up. $125 million dollars down the drain, all because of some bad data.

Unfortunately, like the poor, bad data are always with us. The famous Belgian social statistician Adolphe Quetelet complained in 1846 that medical data were usually 'incomplete, incomparable, suspected, heaped up pell-mell ... and nearly always it is neglected to inquire whether the number of observations is sufficient to inspire confidence'. Likewise, Lord Kelvin was all too aware of the importance of accuracy in measurement. In his Presidential Address to the British Association for the Advancement of Science in 1871, he said: 'nearly all the grandest discoveries of science have been but the rewards of accurate measurement and long-continued labour in the minute sifting of numerical results'.

Even the great Isaac Newton was not immune from bad data: he nearly failed to discover the law of gravity because of it. When he was trying to formulate his theory, he used the best estimates of the Earth's dimensions that he knew, but these were about fifteen per cent too small. As a consequence, the Moon's orbit and the fall of an apple towards the Earth appeared not to follow the same laws. It was only when more accurate measurements of the Earth's dimensions were made that his equations fell into place.

Newton's laws also predicted that the orbit of the planet Mercury should change slightly over time (should *precess*) due to the influence of the other planets, but, no matter how many times the calculations were performed, the calculated value did not exactly match the observed value. The difference between the two values was about $0.012°$ of arc (try measuring that with your protractor and see the impact of measurement error on your results!). Much later, in the twentieth century, Einstein's general theory of relativity predicted exactly this discrepancy. Not all anomalous data are bad.

If these examples of the importance of data accuracy in scientific investigations leave you unmoved, then consider the survey of the Top 500 corporations carried out by PricewaterhouseCoopers which found that poor quality data caused *most* of these corporations to suffer significant problems. What about David Loshin's comment that 'scrap and rework attributable to poor data quality accounts for 20–25 per cent of an organization's budget'. And then there's the number given by the Data Warehouse Institute, which estimated that current data quality problems cost US businessmen more than $600 billion a year. Gross that up across the world, and it is a miracle that we achieve anything at all in the face of poor data!

It's bad enough that poor data can affect the biggest corporations, but it gets even worse: it can affect entire countries. A report by Gabriel Rozenberg, economics reporter of the *Times*, on 2 December 2004 said:

Britain's building industry was plunged into a bogus slump last year by a computer glitch that deleted 66,500 firms from a list used to compile official estimates of construction output ... The data now show that construction growth rose 0.5 per cent in the first three months of 2003, rather than falling 2.6 per cent as originally reported. It grew 2.1 per cent in the second quarter, rather than an immodest 5.3 per cent.

Since economic policies are built on such figures, it is easy to see that bad data like this can lead to disaster.

In his Presidential Address to the Royal Statistical Society in 1979, Sir Claus Moser, former Head of the UK's Government Statistical Service, remarked that the Central Statistical Office had devised a motto to the effect that 'any figure that looks interesting is probably wrong'.

Whether data are bad or not depends very much on the context. The plot of the week-by-week proportion of a bank's loan customers who repaid by direct debit, over a period of about four-and-a-half years, shown in figure 5.1 at the end of the previous chapter, has a very clear step change about halfway through the period. A statistical model based on customer behaviour in the early part of this period would be a very poor model for the second part. The data would be poor quality from this perspective: it would not be representative of the population one wished to describe. And yet there are no errors in the data. Nothing is missing and there is no measurement inaccuracy. It is simply that the behaviour of the customers differs between the two time periods.

Whether data are bad or not is also a function of how they are treated. Coding missing values for age using 99999 is perfectly fine, that is, until a careless analysis includes these values when calculating the average age. The quality of data may thus depend on the interaction between the actual numerical values in the computer and what is done with them.

In fact, one can generalize even beyond the observation that data quality depends on context and intended use. Thus, for example, the International Standard ISO 8402 defines quality as 'the totality of characteristics of an entity that bear on its ability to satisfy stated and implied needs'.

There are many reasons why data may not be as perfect as one would like. Later sections of this chapter describe some of these reasons, but the discussion of one special reason is deferred to Chapter 7. This is the deliberate distortion of data: fraud, cheating, lying, deception in general. That chapter also discusses how perfectly accurate data may be mistreated, or analysed in deliberately misleading ways to give misleading results or to bolster incorrect conclusions. This sort of thing occurs all too often, so it is as well to be aware that it does happen, and to be alert for it.

Accuracy, precision and bias

At first glance, one might assume that the only aspect of data that matters for drawing valid conclusions is that it should be 'accurate'. Accuracy means that the recorded value is close to the true value and this is certainly a fundamental aspect of data quality. However, closer examination shows that accuracy alone is not sufficient. To be useful, data must also be *relevant, timely, accessible, clear, complete* and *consistent*. We look more closely at some of these broader issues of data quality later.

Accuracy itself has two quite distinct aspects: *precision* and *bias*. If a measurement of the same thing under effectively identical conditions is repeated, in an ideal world we would hope that the results would be the same. In reality, of course, there is often some variation in the results, possibly due to random measurement error or other uncontrolled sources of variation. For example, if we repeatedly measure someone's height, we might find a difference of a millimetre or two each time, perhaps because of slight changes in their posture, or because of the way the joints settle due to gravity over the course of the day, or perhaps because of changes attributable to age, or because of slight variation in where we judge the top of their head to be and so on. That is, the repetitions lead to a *distribution* of values. Most of the observed values cluster around some central value, with the more extreme values, either larger or smaller, occurring less often. The further out one goes, the less often the values occur: if someone's 'true height' is 180 cm, then we would expect to get measurements of 180.1 cm quite often, and we would expect to observe measurements of 180.2 cm less often, and very rarely would be expect the result to be 181 cm. We see from this example that it is often convenient to think of this as being a distribution of observed values about the 'true value', with the true value being the 'centre' of the distribution. The arithmetic mean or average discussed in Chapter 5 is often used as a central value. It thus provides an estimate of the true value. The extent of the deviation from this central value is the precision of the measurement procedure. Formal statistical measures of this deviation are widely used to measure precision. For example, a very popular measure is the average of the squared differences between the observed values and the central value – the *variance*, which we also met in Chapter 5.

Of course, it is not merely measurement which leads to a distribution of values. Slight changes in the manufacturing conditions of an object will lead to the objects having slightly different dimensions. Plastic automobile parts, such as fenders, dashboards and so on, are made by an injection-moulding process of the kind mentioned in the context of experimental design in Chapter 5. This process is controlled by a number of variables, including the pressure and temperature of the plastic in the mould, the rate at which the mould is cooled and the precise constituents of the plastic. By varying these, one obtains different quality products. If the cooling is too rapid, for example, then perhaps the plastic will shrink too much. These sorts of ideas (if not in the context of plastic injection-moulding) became particularly important with the Industrial Revolution and later, with the ideas of mass production and exchangeable parts. Only if a part was manufactured with high precision could it be simply plugged in to replace a defective part. If the precision was low, machines would often not fit together properly, and they would not work.

Some modern quality control systems, such as the Taguchi system, have placed great emphasis on the precision of manufactured products. They note that it is one thing to be sure that, on average, the parts have the right dimensions, but if the spread of values is too great it could mean that very few parts are near enough to the desired value to be useful.

Other terms are often used in place of the word precision, with same meaning. These include reproducibility, repeatability and consistency.

Precision, then, describes the width of the distribution of values about the true or intended value. The smaller this spread, the better, since it means that each time a measurement is taken (or a part produced) it will have a value near to the true (or intended) value. In an ideal (and unrealistic) world, the distribution would have no spread.

The other important aspect to accuracy is *bias*, which was also introduced in Chapter 5. To illustrate with an example with which many readers will be familiar, suppose that we wish to measure our weight using bathroom weighing scales. Repeated measurements (stepping onto the scales several times) may show that the measurements are very precise: perhaps they cluster very closely about a value of 70 kg, with most extreme deviations from this value being only a matter of some small number of grams (perhaps attributable to precisely where on the machine you stand, or due to some stickiness in the machinery, which changes as you repeatedly use it, and so on). Then, at least as far as precision goes, these data are accurate. Now suppose that we note that when we step off the scales it still reads 5 kg. It seems as if the device is miscalibrated so that it reads 5 kg higher than it should. Our precise estimate of 70 kg is then all very well, but the data are *biased* by the amount of 5 kg. Our true weight is 65 kg, not 70 kg. Bias, then, tells us about consistent distortion. Even if the scales had perfect precision, so that they gave exactly the same result each time a measurement was taken, if they were biased they would not be accurate: we would be unable to trust the results. More generally, measuring devices do not have perfect precision, so that they produce a distribution of values. Bias tells us the difference between the true value and the average value of this distribution.

Precision and bias are quite distinct from each other. Data can be precise (little variation about the true value) while still being biased (consistent departures from the true value), and they can be unbiased (no consistent departure) while also being imprecise (substantial variation about the true value). To be accurate, data must be both precise and have little bias.

While accuracy is important, one can sometimes be misled. There is such a thing as spurious accuracy. One often witnesses this in the output of computer programs. For example, one might see that the average weight of a group of ten patients is calculated to be 65.897665473 kg. This is clearly ludicrous. While the number might be an accurate value for the mean values of the *recorded numbers* representing the weights, the last few decimal places certainly do not reflect any reality. The three at the end of this figure corresponds to a millionth of a gram. People's weight changes by a kilogram or two overnight simply through loss of liquid exhaled while breathing. Moreover, the original ten measurements from which this figure was calculated may well have been rounded to the nearest tenth of a kilogram, or even the nearest kilogram. This will mean that all but the first few digits in this number represent mere random variation, and yet one sees this sort of thing time and again. Think how you would react if you read that 'there are 6,321,544,443 humans alive on the planet'. Even while you read that sentence, the number changed as people were born and died.

A similar spurious accuracy sometimes arises in newspaper reports when the units of measurement are changed. In a letter to the *Times* on 6 January 2003 commenting on a report of ice blocks 10 m 'or 32.8 ft' high, Mr Dudley Smith drew attention to such absurdities by asking 'would it not have been better to have expressed this as 32 feet nine and five eighths inches, for the benefit of those of us not conversant with decimal feet?'

Distorted data

Distorted data do not properly reflect the object or process about which one wants to collect data. After all, if the aim is to understand something, then data are only good to the extent that they shed light on that something: if I wanted to know the average weight of the players in a football team, it would not help if each weight had a random number of pounds subtracted from it. Data can become distorted in an infinite number of ways, but a useful dichotomy is into those arising from instrumental error and those arising from human error.

Instrumental error arises when there is some fault with the measuring device collecting the data. Miscalibration of the kind referred to in the preceding section might be regarded as a kind of instrumental error. Measuring instruments might be initially miscalibrated (e.g. the zero point incorrectly set) or they might drift out of calibration with time or use. A spring might become less elastic over the course of time or abrasion and wear might degrade fine instruments. Musical instruments drift out of calibration – which is why they need to be retuned occasionally. An unusual example of a calibration problem was discovered by one of my postgraduate students. He noticed that an instrument measuring windspeed occasionally gave excessively large readings. Independent checking showed that the gales this instrument was apparently reporting did not exist. Close examination revealed that the instrument automatically reset itself every midnight, and sometimes when this happened it recorded a very large value on the trace.

Often things being measured have to be prepared in various ways before they are suitable for measurement. Cells may have to be dyed, objects may have to be physically cleaned, patients might have to be carefully positioned for X-rays. If any of these things are not done properly, the measurement will not be measuring what it should. If one regards the preparation as part of the measurement process, then the data are subject to an instrumental distortion.

I have already referred to the new science of proteomics, one of those very exciting areas which promises to revolutionise medicine. In 2002, a paper in the prestigious medical journal *The Lancet* reported the discovery of patterns in proteomic spectra that could distinguish between women with ovarian cancer and those without. Since ovarian cancer is often fatal, the potential of this discovery for screening purposes was very exciting. Unfortunately, further close examination by Keith Baggerly and his co-workers at the Anderson Cancer Center in Houston suggested that the observed difference might not be entirely due to underlying biological differences, but partly due to differences in the ways the samples were treated. The recorded data were not what one thought, but were data that had been distorted by the measurement process.

If proteomics is frontier science, consider this example of centuries old science, and how accurate data had a major impact on civilization. The Earth takes 24 hours to rotate, so one hour corresponds to a rotation of 1/24th of 360°; that is, to 15°. This yields a way of converting from differences in time to differences in longitude. If a navigator carries two clocks, one fixed to mark the time of his home port, and one set to the local time (by setting it to read noon when the sun is at its highest point), then one can convert the difference in time between the two clocks to a difference in longitude between the two positions. If the clocks read one hour differently, then they are 15° apart, two hours corresponds to 30°, etc.

This is all very well, but it obviously hinges on having accurate clocks. Indeed, not merely accurate clocks, but clocks which are unaffected by the rigours of ocean travel: the constant swaying, the unpredictable batterings of storms, the salt water, the changes in temperature and humidity, etc. To enable navigators to know their longitude, none of these things must knock the clock out of calibration. Nowadays, of course, this is no problem: everyday wristwatches and electronic clocks, even radio clocks tuned to international standards of time, allow one to know the time to any degree of accuracy one could wish for. Three centuries ago, when exploration of the seas was at its height, things were very different.

The story of the attempt to build an accurate clock which could function in ocean-going ships is a protracted one, extending over centuries. It involves princes and kings, huge sums of money (both sums lost due to inaccurate clocks incorrectly giving the longitude, and sums offered as prize money to anyone who could develop an accurate clock), the loss of countless sailors' lives, and also the greatest of scientific minds. Galileo, Huygens, Newton, Halley and any number of others spent time trying to resolve the problem of how to accurately tell time at sea. It also involved many personal disputes and enmities.

Eventually the English clockmaker, John Harrison, having devoted his life to the problem, managed to crack it. He abandoned the use of the pendulum, which was the most common tool for making each second the same length as the others. This removed, or at least eased, problems due to the rocking of the ship. Harrison was an expert on the metals used in clocks, and he combined these in such a way that when one expanded due to a change in temperature, another would contract, so that the whole system remained constant. Now the clock would not speed up or slow down as the ship moved from colder to warmer climates. Instrumental error would not intrude, so that sailors could know exactly where they were.

Various types of error occur so frequently that they are given names of their own. For example, the *ceiling effect* and *floor effect*. These arise when instruments have restrictions on the range of values they can record, with maximum or minimum possible recorded values. A mercury thermometer, for example, would be of limited value for measuring temperatures lower than $-40\,^\circ$Celsius. This is the freezing point of mercury, so the instrument would show a floor effect, yielding many reports of a temperature of -40° and none lower than this.

The 'instruments' here need not be physical instruments, but can also be social or psychological measuring instruments. For example, the UK system of examinations which are taken at age 18 – the so-called Advanced levels or A-levels – have been subject to a ceiling effect in recent years because of grade inflation. Grade inflation is the tendency for average marks to creep up over time, not necessarily because the students are getting cleverer, but for more complicated socio-economic reasons. Perhaps the shift from examinations to assessed course-work has contributed. Perhaps it is the fact that there are several different boards setting their own examinations, so that they will be competing to encourage more students to sit their examinations (and pay the fee). Naturally students will prefer to pick an exam board which is most likely to award them a high grade. The result: a gradual increase in the number of high grades awarded. The consequence is a ceiling effect, as more and more students get the top grade, and so cannot be distinguished.

Like instrumental error, human recording error can also occur in an unlimited number of ways. One can misread a scale, transcribe values incorrectly, misplace decimal points, read or write down the wrong figures, write down illegible figures, misunderstand questions, record the wrong units, and so on. Sometimes poor human choice causes problems. Coding missing values for age using 99 is obviously not an ideal choice – though this has not prevented it from being done.

Such errors may not occur too often, but a small rate of incidence means one is almost certain to find them in large collections of data. William Kruskal, in his 1978 Fisher Memorial Lecture, described data from the 1962 US Population Census apparently showing that 62 young women aged 15 through 19 had 12 or more children. As he put it:

> Although I suppose it barely possible physiologically that a 19-year-old woman have 12 or more children, in our society it is unthinkable to me that there be 62 such. Some possible explanations are incorrect age or number of children given by the respondent, errors of transcription, coding, and so on. For the respondent errors one can imagine many scenarios. For example, a mother providing information about an adult daughter may confusedly ascribe to the daughter the mother's number of children.

Kruskal also gives other examples of recording error in census data. For example, the 1950 US Census contained an excessive number of 14-year-old widowers. Close examination revealed that this had arisen from translating the digits one space to the right on punched cards. This was not a one-off: the 1970 US Census showed that there were apparently 2,983 14-year-old widowers, and 289 young women who had been both widowed and divorced by 14. This

Census also showed that there were apparently 2,926 males aged 25 to 29 enrolled in the first grade.

In fact, based on a lifetime of experience in analysing data, Kruskal remarked: 'A reasonably perceptive person, with some common sense and a head for figures, can sit down with almost any structured and substantial data set or statistical compilation and find strange-looking numbers in less than an hour.' Bad data, or at least suspicious data, are always with us.

Units of measurement are a potential trap for the unwary. In medical records, even though the units in which measurements should be taken are always specified, people sometimes fail to notice this. So we get height given (or requested) in feet and in centimetres, and weight in pounds and kilograms. This is no problem if the units are noted – but it becomes a problem if the required units are kilograms and those recorded are in pounds without an indication that this is the case. Too much of this could lead to an apparent epidemic of obesity. And this sort of thing does happen. In recording birth weights, measurements might be taken in ounces but written down as pounds, sometimes the symbol for the unit of the pound, lb, is misread (for example, reading 1lb as 11lb), misplaced decimal points lead to birth weights being ten times their real value, and so on. While instrumental errors might often be expected to lead to biased figures (e.g., a broken instrument always recording 0), many of these human recording errors of birth weights also tend to lead to errors in the same direction, and it has been suggested that this could account for an observed apparent excess of overweight young babies.

An error involving units occurred in *The Times* on 19 December 2002. A short report noted that 'Britain's largest bat, the greater mouse-eared bat, which was officially declared extinct in the UK 12 years ago, has been rediscovered hibernating in an underground hole in West Sussex. They can weigh up to 30 kg and have ears as long as 3 cm.' The next day *The Times* carried several letters commenting on this discovery. Mr Tim Bloomfield remarked that '30 kg, the weight you quote, is a lot of bat – more like several large turkeys, or my seven-year-old nephew in his Batman suit. Have you any idea of the collateral damage a single 30 kg bat could do if it recklessly flew into you one night? Or the type of diet required to sustain such a remarkable animal?' Mr Bill Cairns commented 'If the greater mouse-eared bat weighs some 66 lb, as a resident of West Sussex I feel I should stay indoors at night until it is declared extinct again.' Mrs Janis Mason said 'Why worry about possible terrorist attacks? Should we not be more concerned about the dangers of meeting with a 30 kg bat when it comes out of hibernation?' Mr Godfrey Curtis set it in context: 'Monty, our somewhat overweight yellow labrador, just about weighs 30 kg on a good day ... The thought of him equipped with leathery wings and airborne fair boggles the imagination,' as did Mr Derek Cannell: 'No wonder the mouse-eared bat is hiding in a hole in the ground ... It is probably waiting for the results of the airport inquiry to see if there will be a runway big enough to use for take-off.'

The newspaper followed up these letters with a footnote saying that a greater mouse-eared bat usually weighs about 30 g, not 30 kg.

I remarked above that sometimes particular types of errors occur so frequently that they are given a name of their own. 'Digit preference' or 'heaping' is one such common data-recording distortion phenomenon familiar to statisticians. It describes the subconscious tendency for people to round values to the nearest convenient number. A beautiful example of this concerning blood pressure measurements was pointed out to me by a former colleague, Dr Fergus Daly. The data give the diastolic blood pressure readings, in millimetres of mercury (mm), of around ten thousand men. Examination of a plot of the data showed very pronounced peaks: there were far more observations at 60, 70 and 80 mm than at neighbouring values. Now a priori it seems unlikely that nature has contrived the human blood circulatory system so that there is an abundance of men with blood pressures at such convenient round numbers. It seems much more likely that some peculiarity of the data collection process has led to it. In particular, it seems likely that the digit preference phenomenon accounts for it, whereby the people taking the measurements have rounded things to the nearest 10 mm of mercury.

As it happens, these data also contained some other oddities. For example, below about 80 mm, there were very few odd-numbered values recorded, and where odd-numbered values did arise, they all ended in 5. For example, there were no recorded values of 61, 63, 67, or 69 mm. Now it is difficult to imagine a mechanism in the human body which would lead to many people having diastolic blood pressure values of 60 mm, but none having values of 59 mm or 61 mm. Again, it seems much more likely that some peculiarity of the data collection procedure accounted for it. And this, in fact, appeared to be the case. First, the instrument used for measuring blood pressure (a *sphygmomanometer*) was graduated in only even units, and clearly nurses had a strong preference to take the instrumental values as the true readings. Except, that is, when the values were centrally located at around 55 mm, 65 mm and 75 mm. Here the same digit preference phenomenon probably accounted for the small but distinct numbers of values recorded as ending in 5. As to the fact that all sorts of odd numbers were recorded for larger blood pressure values, this is had apparently arisen because, when an anomalously high value was recorded, the nurse was asked to repeat the measurement and take the average of the two results. Even something as simple as measuring blood pressure has hidden complexities, and can lead to distorted data.

Roberts and Brewer (2001), in a paper in the *Journal of Applied Statistics*, describe another example of digit preference arising in a study of drug usage amongst persons at high risk of HIV. When asked to report the number of drug partners, they reported exceptionally large counts at ten, twenty and thirty, and also anomalously high values at fifteen and twenty-five. Interestingly enough, when asked to think of them by name, and count the number, the exceptionally high values vanished, being smoothed into the general distribution of numbers.

One must be wary of leaping to conclusions about the causes of particular kinds of anomalies in a data set. Digit preference is not the only type of recording error that can lead to surprising peaks in a data set. My former graduate student Gordon Blunt discovered some striking errors in a collection of figures on insurance premiums: two very large peaks at relatively small values. Closer investigation revealed that they arose due to miscodings in the way the transactions were coded up for analysis.

Rounding to the nearest whole (or convenient) number can sometimes lead to incorrect conclusions. In the nineteenth century, the French statistician Adolphe Bertillon discovered that the distribution of heights of 9002 military conscripts in eastern France did not follow the expected symmetric distribution with a single peak, but had two peaks. On the basis of this, he conjectured that the population in that region was comprised of two different racial groups – and it was later found that indeed this was the case, with the local population being a mix of Celts and Burgundians. In fact, however, this was a case of distorted data leading to the *correct* conclusion. Ridolfo Livi, at the end of the nineteenth century, looked at things more closely. He discovered that the dual peak had arisen purely because of the way the data had been treated. In particular, the heights were originally rounded to the nearest centimetre, but Bertillon had converted them to inches, which he had then further rounded to the nearest inch. This had the unusual and unexpected effect of putting three of the original one centimetre cell counts into each of the one inch cell counts between 4'11" and 5'4", *except* the one inch cell count between 5'1" and 5'2", which had only two counts. The result was a distribution with two peaks.

Of course, anomalous and large peaks in data sets can also arise for perfectly sound reasons. Prices of goods in shops often show exceptionally large numbers of items priced near whole numbers of dollars or pounds. This is due to the common practice of pricing things at just less than a whole dollar number: $9.99, $19.99, etc.

Digit preference might be described as a consequence of human psychology: people have a natural tendency to round numbers to convenient values. Other errors arise from simple oversight, and sometimes from the extreme complexity of the systems humans create. In another financial data set containing data about unsecured personal loans, Dr Mark Kelly and I found tiny amounts unpaid (e.g. 1p or 2p) leading to a customer being classified as a 'bad debt', negative values in the amount owed, twelve month loans still active after twenty-four months (technically not possible under the bank's rules), outstanding balances dropping to zero and then becoming positive again, balances which were always zero, and number of months in arrears increasing by more than a single integer in one month. All of these are problems with the data.

One can explain some of them. For example, the bank's 'month' lasted for four weeks, and most months have more than $4 \times 7 = 28$ days in them. The bank also managed to explain other anomalies when we brought them to its attention. However, they could not explain all of the problems. Some of them were doubtless caused by systems put in place long ago, with the people who originally coded them into the software now long gone, along with their rationale for doing so. With those examples fresh in your mind, how confident are you now that your bank has your transaction details correct?

Digit preference is an example of how subconscious psychological effects can lead to distorted data. Another, and a rather more complicated example, has been termed the Hawthorne effect. This describes the apparent psychological phenomenon that the mere fact of being the subject of an investigation can induce a change in the behaviour of a subject. The original studies were carried out at Western Electric Company's Hawthorne Works in Cicero, Chicago, around 1930. The original aim of the study, at the beginning of 'industrial psychology', was to explore how various changes to the working conditions led to improved productivity amongst a group of workers assembling telephone relays. The researchers found it very difficult to pin down what caused productivity changes. Altering the level of illumination appeared to have no direct link with level of productivity. Likewise, changing the incentive scheme and altering the time allowed for breaks had no obvious correlation with output (although, for example, introducing six five-minute breaks led to a drop in productivity, presumably because of the interruption these cause). However, it was noticeable that there was a general trend of increasing productivity over the experiment: whatever intervention was tried, broadly speaking, things improved, even when previous conditions were resumed.

The researchers proposed several hypotheses for these startling results, but dismissed all but one after further thought. What they were left with was the suggestion that *the presence of an observer throughout the study had led to the increase in productivity*. This observer did not passively observe, but also interacted with the group, informing them about the study and also actively requesting their views. The presence of the observer in this role showed the workers that someone was interested in them, that they were the subject of special attention, and the suggestion is that it was this that caused the improvement. It is this phenomenon, that the mere fact of (knowingly) being a subject in a study leads to a change in behaviour, which is termed the Hawthorne effect.

As with deliberate distortion, the data resulting from a study such as this are 'bad' in a sense different from most of the cases described above. They are bad because they do not measure the thing one wanted to measure: the effect of altering the working conditions. Instead some other factor has intruded to change the data. They are 'good' in the sense that they accurately record the effect induced by this extra factor. 'Badness' or 'goodness' of data depends on the context, on what one intends to do with the data, and what questions one hopes to answer.

The Hawthorne effect has become a staple of industrial psychology, and an accepted scientific 'truth'. In fact, this is rather ironic, because the data on which it is based, the data collected on the relay assemblers in the Hawthorne Works in Cicero, is bad in a more direct sense as well. It seems that the management style, which started out as supportive and friendly to the workers, reacted to the way they were behaving. Initially encouraging the ten workers to work in a relaxed way, when productivity began to fall the managers replaced the two of the workers who seemed to be working the least hard halfway through the study. One of the replacements was very highly motivated to make as much as she could from the performance-related pay scheme (her father had recently lost his job).

Since, moreover, this scheme depended on the overall group output, and not on individual output, she was also very highly motivated to encourage the others to work hard. Thus, these changes to the constitution of the group almost certainly had a major impact on productivity. Put bluntly, there were major inadequacies in the design and conduct of the experiment, which casts doubt on its conclusions.

Of course, the inadequacies of the original study do not mean that the Hawthorne effect does not exist. It has obvious similarities to the *placebo effect*. This is a well-attested medical phenomenon which certainly exists. The placebo effect arises when, unknown to them, patients are given an inactive 'treatment': flavoured water in place of medicine, for example. This is done in clinical trials where one is testing the effect of a proposed new treatment. As we saw in Chapter 5, in such cases it is crucial to have a control group with which to compare the treatment group, so that one can be sure that any improvement is not due to some factor other than the treatment. Thus, in the case of the common cold, people just naturally recover over the course of time. Without a control group receiving no treatment to compare with the group receiving the active treatment, one might be tempted to attribute the recovery to the medicine. The fact is, however, that studies have shown that often, despite the fact that there can be no biological mechanism leading to improvement, patients often record a small improvement when given a nonactive control, or placebo treatment (they, of course, are not told that it is inactive). It seems likely that the effect is a psychosomatic one: the mind believes that it is being given an active treatment, and this has some beneficial effect. Of course, the effect is generally small, and typically does not persist in time: placebo treatments typically do not cure organic illnesses. Homeopathists should take note of this. In homeopathy the 'treatment' is so diluted as to be non-existent, and any apparently beneficial effects of homeopathy are likely to be purely because of the attention the therapist gives to the patient (which is often more than can be given by time-pressured conventional doctors) coupled with the placebo effect.

The Hawthorne effect and the placebo effect involve a subtle interaction between the subject being studied and the measurement
procedure, serving to distort the results, so yielding data that is 'bad'
in the sense that it does not describe what you want described. Similar
effects occur more generally, as feedback mechanisms, especially
when human beings are the subject of study. Then, often, attention
focuses narrowly on the particular measure being used, to the extent
that this becomes poor as an indicator of what is really of interest. For
example, in the context of medicine, the infant mortality rate is often
used as a general measure of population health for less developed
countries. Unfortunately, as a consequence this particular measure is
targeted for improvement, so that it becomes less and less useful as a
general measure of population health. In the context of economics,
this sort of phenomenon has been called Goodhart's Law (Charles
Goodhart was Chief Adviser to the Bank of England), and has been
expressed in various ways, including 'Any observed statistical regular-
ity will tend to collapse once pressure is placed upon it for control purposes' (Goodhart, 1984) 'As soon as the government attempts to
regulate any particular set of financial assets, these become unreliable
as indicators of economic trends' (Pears Cyclopedia, 1990), and
'Performance indicators lose their effectiveness when used for policy
decisions' (Strathern, 1997). Economics is a slippery science!

The fundamental issue is, as we have already noted, that it is not
possible to measure something without interacting with it in some
way. In many situations this is not an issue, because the interaction
leads to infinitesimal changes in the thing being studied. In other
cases, careful experimental design is needed to get round it. I
referred to this in chapter 5: we can use randomized response so that
we do not know the answer for any particular respondent, or we can
use concealed cameras, etc. In subatomic physics, however, things
are more difficult still: in some cases it is not possible, no matter how
clever and subtle one's measurement procedures, to avoid a measurement changing the system being measured. This is not a question
of careless design, but rather is a fundamental aspect of the way the
universe appears to work. Once again, the data arising from such

measurements are not bad in the sense of incorrectly representing the underlying physical process: they are only bad in the sense that they do not reflect the process which exists immediately after the measurement has been taken.

When we looked at the Hawthorne effect, we saw that problems arose because of poor design or conduct of the experiment. This is an all too common – and nowadays essentially unforgivable – cause of bad data. A classic example occurred in the UK around 1930. This was the Lanarkshire milk study.

In 1930, a major study was conducted to explore the beneficial effects of adding milk to the diets of schoolchildren: 10,000 school children aged five–twelve inclusive were given three-quarters of a pint of milk a day, with half being given raw milk, and half pasteurised milk. Another 10,000 were given no milk. All 20,000 had their weight and height measured at the beginning and end of the experiment. As you might imagine, all this took some organising. In all, sixty-seven schools were involved, with between 200 and 400 children in each chosen for the experiment. The researchers took great care to see that the study was properly designed, and in each school half of the children were given milk, and half not. Great care was also taken over the measurements:

> All of the children were weighed without their boots or shoes and wearing only their ordinary indoor clothing. The boys were made to turn out the miscellaneous collection of articles which is normally found in their pockets, and overcoats, mufflers, etc., were also discarded. Where a child was found to be wearing three or four jerseys – a not uncommon experience – all in excess of one were removed.
>
> To obviate any slight variation that might exist in the various weighing machines, the same machine was employed at each school both for the initial and final weighing ... A medical officer and a nurse, or two nurses, formed a 'team,' and each team was allocated its group of schools, which were visited in rotation. The same rotation was observed when the final measurements came to be taken at the close of the investigation.

Likewise, great care was taken regarding the quality and homogeneity of the milk, and the time of day it was given to the children.

I shall let Gerald Leighton (Medical Officer – Foods) and Peter McKinlay (Medical Officer – Statistics), who prepared the report on the project, pick up the story:

> Doubtless owing to the inaugural meeting at Hamilton when the Under Secretary of State for Scotland explained the scheme to the public, very widespread interest was aroused at the beginning in the whole district. Some of the parents of the children, however, did not at first grasp what was actually going to be done, or the objects in view.
>
> Dr Macintyre says:
>
> Some resented what they chose to regard as a charity, others that their children had been singled out as ill-nourished and in need of additional sustenance. There were others, again, who objected to their children being 'experimented upon'.

The attitude of the children was also very interesting:

> At the commencement of the test, when the selection of the 'feeders' and 'controls' had to be made, there was almost universal regret that all could not be included in the former category.
>
> When, however, the scheme was fully explained to the children, the 'drys' took the decision in a sporting spirit and concealed their chagrin very successfully. On the other hand the 'wets' did not show any marked desire to exult over their successes in the ballot; but rather seemed to evince a sincere sympathy for those who had been unsuccessful.

Leighton and McKinlay's report also describes some personal impressions given by the teachers involved in the study. These are quite moving, describing the improvement in 'the bloom of [the feeders'] cheeks and the sleekness of their skins,' 'in the playground buoyancy and pugnacity are developing to an alarming extent,' 'mental lassitude gave place to alertness, especially among the younger children'.

Unfortunately, despite the researchers' best efforts, the study was not perfect. Leighton and McKinlay again:

The selection of the actual children, however, was left to the head teacher in each school, to whom it was explained that the selected children should be a representative group of all, and not the most ill-nourished or of any other outstanding character. In the same way it was explained to the teachers that the 'controls' should also be representative of the average child. Likewise it was laid down that the sexes should be, as far as possible, balanced in each age group.

As a matter of fact, the teachers selected the two classes of pupils, those getting milk and those acting as 'controls,' in two different ways. In certain cases they selected them by ballot and in others on an alphabetical system. *In any particular school where there was any group to which these methods had given an undue proportion of well-fed or ill-nourished children, others were substituted in order to attain a more level selection.* (My italics.)

This last point is crucial: it means that the randomization, which we saw in Chapter 5 is necessary to permit valid statistical inference, and which can also materially increase the chances of having representative allocation to the two groups has now been interfered with. In fact, we can see that the study is not even singly blind (the children knew which group they were in), let alone double-blind (the researchers also knew which group each child was in): 'For the purpose of keeping the nutritional record of every child concerned in the investigation, cards were printed having distinctive colours. The children who were getting the raw milk had yellow cards, those getting the pasteurised milk had pink cards, while those for the "controls" were white.'

To assess the effect of the milk supplement, the average initial weights and heights for each age-by-sex group were calculated. It is particularly interesting, and revealing, to note that 'The control groups compared with the two milk-fed groups were, at the beginning of the experiment, slightly, though in many instances, significantly heavier and taller.' This is highly suspicious, especially since 'the differences in weight and height of the two milk-fed groups are smaller, generally insignificant and inconstant in sign'. It seems likely that the 'adjustments' permitted to the initial random allocations have caused this difference: there has been a tendency, probably subconscious, for some teachers to allocate the smaller and more needy pupils to the 'feeder' group.

There were also other complications regarding the weights. For example, although efforts were made to remove superfluous clothing, some adjustment is needed for the fact that the winter clothes are likely to be heavier than the summer clothes. By itself, this would not matter: had the groups been randomly assigned, it would have affected them both equally. But, as we have seen, they were not, and it is likely that the size of adjustment should be different for poorer and richer pupils.

The report concludes: 'The influence of the addition of milk to the diet of school children is reflected in a definite increase in the rate of growth both in height and weight.' In a critical assessment of the study published in 1931, William Sealy Gossett (1867–1937) who wrote under the pen name of Student since his employer, Guinness, would not allow him to use his real name, commented:

> This conclusion was probably true; the average increase for boys' and girls' height was 8 per cent and 10 per cent over 'controls' and for boys' and girls' weights was 30 per cent and 45 per cent, respectively, and though, as pointed out, the figures for weights were wholly unreliable it is likely enough that a substantial part of the difference in height and a small part of that in weight were really due to the good effect of the milk. The conclusion is, however, shifted from the sure ground of scientific inference to the less satisfactory foundation of mere authority and guesswork by the fact that the 'controls' and 'feeders' were not randomly selected.

Most of the examples above have arisen because of problems with individual values or recordings. Other problems involve entire databases. Indeed, the problems with the Lanarkshire milk study might be regarded as of this kind, since the distortions affected entire groups. Poor *record linkage* is another common cause of data errors which affects entire databases. Record linkage is the process of merging two or more databases, and is becoming increasingly important in the commercial world, where the separate operational databases used for the day-to-day running of the company (e.g. one for marketing purposes, another for risk assessment of applicants, another for monitoring of existing customers, and so on) are combined into a single giant data warehouse, which is used for data mining to discover the properties and behaviour patterns of people.

In fact, the author experienced the sharp end of this, albeit on a smaller scale, and fortunately without any serious consequences. A few years ago I had occasion to go through all my medical records held by my doctor. I discovered that I was recorded as having had a tonsillectomy when I was a child. I had no recollection of this, and a swift examination showed that either my tonsils had regrown, or they had in fact never been removed. Further detective work revealed that the records of two people with the same name had been combined. It took considerable work to pick them apart.

I referred to grade inflation above. Grade inflation is an example of something changing over the course of time. In this case, it is the distribution of students' examination scores changing, but change can occur in many ways. If one is not aware of the changes, if one does not make allowance for them, then the data become distorted for practical purposes. For example, suppose one constructed a medical screening system to screen for cancer. Naturally, over the course of time, human populations change: people move into and out of an area, people from different racial backgrounds, with different tendencies to develop cancer may come or go, diets, known to influence the incidence of cancer, change over time, population age structures change and the nature of diseases themselves change (think of flu epidemics which sweep the world and then die out). All of these things inevitably mean that a screening system will gradually become out of date. The data on which it was originally constructed are 'bad', in the sense that they no longer reflect the people on whom they are going to be used.

Fortunately, such problems are rare in medical contexts: human populations change relatively slowly – more slowly than changes in medical technology. The same is not true, however, for populations of customers of banks. Banks use scorecards to decide to whom to lend money. These scorecards are typically based on the records of previous customers. So, for example, suppose we wanted to construct a scorecard to decide to whom to lend money for a two year period. We could look at our past records of customers who had received loans in the past, identify those who repaid on time and those who didn't, and then build a statistical model to identify in what way these two groups differed in terms of the characteristics on their application forms (income, time at present address, repayment record on credit card bills, etc.). This is fine so far as it goes, but now consider the same idea applied to a twenty-five-year mortgage. If we want to look back in time to see which previous customers turned out to be good or bad, we need to look back twenty-five years or more. It seems highly likely that the characteristics and behaviour patterns of customers from that long ago will bear very little resemblance to those of today. The nature of the populations will have changed, society will have changed, certainly the economy has changed. Indeed, the world has changed. Any model built on data which are twenty-five years out of date will probably be of very limited value for predicting modern day customers. In this sense, the data are bad.

Missing data

Missing data are a frequent occurrence (or perhaps that should read 'are a frequent non-occurrence'). Data can be missing for a huge variety of reasons: if the data are about people, then perhaps subjects or customers simply refused to answer; if the data are physical observations, then perhaps the instrument broke down; sometimes it is logically impossible for data to be observed (your spouse's income, if you are not married) etc. An important distinction is between items missing from records (e.g. someone who answers all the questions, except stating their age), and situations when entire records are missing. The latter can be particularly problematic, since one may not know that they are missing. These two situations merit separate examination.

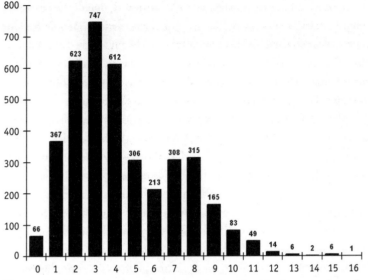

Figure 6.1 Numbers of applicants (vertical axis) with given number of missing values (horizontal axis) on a twenty-five-item application form amongst 3884 applicants for bank loans.

Just how serious the first type can be, when all records are present but not all values in all records are present, is illustrated by the example in figure 6.1. This shows a bar chart of the numbers of missing answers from twenty-five questions given by 3884 applicants for bank loans. The chart shows how many applicants had no missing values, one missing value, and so on. We can see that only sixty-six applicants provided complete information. At the other extreme, one applicant had sixteen of the twenty-five values missing (I do not know whether this applicant was granted a loan on the strength of their nine answered questions!). Only five of the variables had no missing values, and two had over 2000 missing values. In fact, in the last sentence things are not as bad as they may seem. Sometimes *structural* missing values occur. These are the logical impossibilities referred to above – inevitable consequences of the form of the data. For example, your salary will be a missing value if you do not have a job.

In one banking scorecard that I encountered, one of the customer characteristics was omitted from the scorecard altogether for one entire year so that, for this year, all the applicants scored consistently low. The bank had not noticed this, and it wasn't until we started to use the data to develop improved scorecards that the problem came to light. This particular bank has since been taken over!

A similar example occurred with a company which made small personal loans to the 'sub-prime' market – the higher risk sector of the community. Analysis of their data showed a gradual reduction in the quality of some aspects of the data over the course of time: things seemed to be getting worse instead of better as time progressed. Investigation inside the company revealed that those who were responsible for entering the data were gradually not bothering to enter those items for which they could not see any immediate reason. And indeed, there was no immediate reason: the reason would only become apparent some months down the line when the data were analysed to try to detect behaviour patterns in the customers. But by then it was too late.

In general, it is not uncommon for *all cases* to have *some* missing items. This means that if one decided to base an analysis on only the complete records, one would have no data at all! Somehow, one has to be able to cope with incomplete records.

When individual fields are missing from records, at least this is obvious – there will be blanks or missing value codes in the data. However, when entire records are missing, the problem may be a great deal more serious: one may not be aware that there are records missing. This sort of problem arises with population censuses, of course, which attempt to involve the entire population of a country but may miss out on large numbers of people (illegal immigrants or others who, for reasons of their own, do not want to appear on any official database). It goes without saying that any conclusions based on a collection of records that are not properly representative of the population being studied may lead to completely misleading conclusions about that population. If I were to try to draw conclusions about the inhabitants of New York on the basis of those who were prepared to spend two hours answering my questions, I would doubtless miss out many people who did not have two hours to kill. I would only have myself to blame for the inadequacy of my results. We looked at similar issues in the section on collecting data (p. 114).

These sorts of problems are not new, and thinkers have been aware of them in the past. Francis Bacon, in his *Novum Organum* (Aphorism XLVI) of 1620 said:

> And therefore it was a good answer that was made by one who, when they showed him hanging in a temple a picture of those who had paid their vows as having escaped shipwreck, and would have him say whether he did not now acknowledge the power of the gods — 'Aye,' asked he again, 'but where are they painted that were drowned after their vows?'

A more modern example arises in road traffic accidents, where serious accidents are reported rigorously and accurately, while less serious ones may not be reported at all. This leads to a distortion in the apparent distribution of accidents. In actuarial work, life insurance rates are based on quantitative evidence about human mortality. Reliable data is fundamental. The trick, then, is to select which lives to include in the calculations, and this was a topic of great concern in the early days of the actuarial profession.

I referred above to the use of scorecards in banking to decide who should be given loans. We have seen that these are statistical models based on previous customers. That sounds fine, until one recognizes that these 'previous customers' were just the ones that were expected to repay the loans on time — after all, one only granted loans to these. They were not a representative sample from the entire population of applicants for loans. This is all very well, but we want our scorecard to apply to the entire population of applicants. The sample we have is distorted (for our purposes), and any model based on it is likely to lead to unreliable conclusions.

Let me give an extreme example. Suppose that it is known that some factor (you can think of it as age, time with current employer, etc. – I'll call it X) is highly related to riskiness. For simplicity, suppose that this factor can take only two values, which I shall label as 0 and 1, and suppose that most people with X value 0 default and very few people with X value 1 default. That is, X is a superb predictor of riskiness. Because it is so good, I naturally do not accept anyone who has a value of 0 on X, but only those who have a value of 1. Two years later, when all the customers have come to the end of their loan period, I look at the outcomes. Most people have repaid the loan, but a very few have not, and I want to use the data I have to build a scorecard to predict riskiness for new applicants. The problem is that all the customers for whom I know the true outcome have a value of 1 on X. None of them have a value of 0 – because I rejected such customers at the start. This means that if I built a scorecard using the data I have, X would not figure in it – there is nothing in my data which shows that X is a good predictor of outcome. At this point, I suppose, the bank I work for goes bust.

Note that, in this example, the absence of the true outcome class for those not granted a loan is part of the data generating process, not an accident of the data collection process. If I did know the outcome class of those rejected (whatever that means – after all, they don't have an outcome class because they were rejected!) I could build a model for the entire population of applicants. There is clearly something of a Catch-22 in all of this.

In fact, this sort of complication arises more widely, and in more complicated situations. In medical screening, for example, one also builds a statistical model to predict the probability that a person will have (or will get) the disease. Then one intervenes and treats those that one thinks will have/get the disease. The aim is to cure/prevent it, so that one can hardly then use the outcome (have they got or will they get the disease?) as a measure of how effective the predictive model is. The very intervention will distort the data. Of course, the data are not really bad, but certainly from the perspective of a model designed to predict the class membership, they are far from ideal.

As a general point, note that any step in processing data, preparing it for processing, transferring it from one system or representation to another, or adjusting it or altering it is a potential place for the introduction of errors or for data to be lost. The more steps that are involved, the greater is the chance of introducing distortion.

In case the examples of missing data above give the impression that the issue may not have serious consequences, here is another example, with an altogether more serious outcome.

On 28 January 1986, the *Challenger* space shuttle blasted off. On the night before the launch, there was a three hour conference call between people from Morton Thiokol, who made the solid rocket boosters, NASA people concerned with motor control design from the Marshall Space Flight Center, and people from the Kennedy Space Center. The discussion hinged around the fact that the temperature forecast for the intended launch time the next morning was 31 °F, and there was concern about the effect of such a low temperature on the O-rings, the sealing devices around the joints of the solid rocket boosters. Much of the discussion hinged around data displayed in figure 6.2. Each point on this plot corresponds to a previous launch and shows the air temperature at the time of the launch (on the horizontal axis) plotted against the number of O-rings that experienced 'distress' in that launch. So, for example, a single launch took place when the air temperature was 53 °F, and that launch had three O-ring incidents, and there was a single launch when the temperature was 57 °F and it had one O-ring incident.

From the data displayed here, it seems that there is essentially no relationship, or at best a very weak one, between air temperature and number of incidents. Despite this, some of the participants in the teleconference urged caution, and recommended that the launch should be postponed until the temperature rose above 53 °F – which was the lowest temperature of any previous launch. However, because the data in the figure showed no apparent relationship the launch went ahead. In retrospect, this was obviously a serious error: the Shuttle exploded soon after it was launched, killing everyone on board.

However, the error was at least one of data analysis as one of the decision. The people present at the meeting had overlooked two

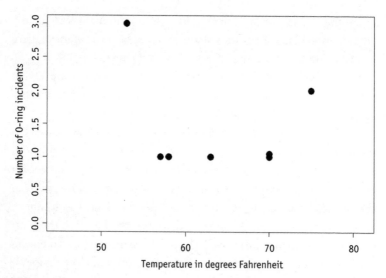

Figure 6.2 Number of O-ring incidents plotted against air temperature.

elementary rules of data analysis: (i) *one should look at all of the data*, and (ii) *one should not extrapolate beyond the data*.

Turning to rule (i) first, figure 6.3 is a plot of the complete data. This shows all of the data in figure 6.2, but also the flights where no O-ring incidents occurred. It is immediately striking from this figure that relatively few incidents occur at higher temperatures, but that proportionately more occur when the temperature is low. There were not very many launches when the air temperature was below 65 °F, but *all of these* are associated with O-ring incidents. In contrast, launches at higher temperatures are associated with very few or even no incidents. This immediately suggests that there is in fact quite a strong relationship between O-ring incident and air temperature, with the risk of an incident increasing quite dramatically as the temperature falls.

On the face of it, it may seem that the launches where there were no incidents can shed no light on the probability that an incident will occur, and this is presumably the view that the meeting took. But this is quite wrong. In this problem, we are trying to make a statement

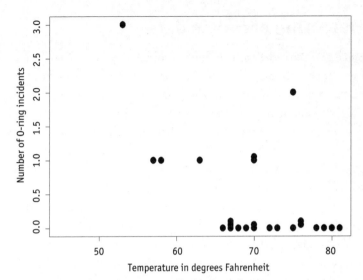

Figure 6.3 Number of O-ring incidents plotted against air temperature: the complete data

about the relative probability of an incident occurring when the temperature is high and when the temperature is low. Launches with 'no incidents' contribute to those estimates of probability. We must be really certain that data are irrelevant before we can afford to ignore it.

The relevance of rule (ii) is perhaps more obvious. With the lowest previous launch temperature being 53 °F, one has very little basis on which to judge the risk of a launch at 31 °F. For all we know, the risk increases dramatically with every degree the temperature falls below 53 °F. Extrapolating into regions where there are no data can be very dangerous.

Of course, it is easy to be wise after the fact. The trick is to learn from one's mistakes. NASA certainly did this – they organized a series of seminars on probabilistic risk assessments throughout the NASA facilities and agreed to undertake 'a comprehensive and more detailed [probabilistic risk assessment] study of the Shuttle throughout all its active mission phases from launch to wheel-stop on landing' (NASA, 2000).

Detecting errors in data

Since data can be bad in an infinite number of ways, we need an infinite number of tests to detect bad data. Put more realistically, no matter how hard we try, no matter how rigorous we are in probing the data looking for anomalies, we can never be 100 per cent confident that there are no problems with the data. Indeed, since we have seen that data may be perfectly fine for one use but not for another, a simple change in what we are doing with the data may change good to bad, without changing its content at all. Furthermore, there will always be errors that are intrinsically undetectable – if, for example, the values could easily have arisen legitimately.

All this is perhaps rather gloomy: we can never be sure our data are perfect. But on the other hand, nothing in life is perfect, so we shouldn't be too depressed about it.

A possible categorization of methods for detecting bad data is into two groups: those which depend on functional relationships, and those which depend on statistical relationships. Let's look at functional relationships first.

Many characteristics have natural bounds. Bounds are values that a characteristic simply cannot exceed, by its very nature. For example, if a characteristic permits only a limited set of categories as a response (e.g. 'does the respondent have a telephone?', with 'no' coded as 0 and 'yes' as 1) then any value outside this set is an error (e.g. a code of 2 in answer to this question is simply wrong). Lower bounds are common: the number of children in a family or the amount someone earns cannot be negative, so there is a lower bound of 0. A reported number of -2 children should immediately arouse suspicion because it does not make sense: there is an error in the data.

Bounds on single characteristics are a particularly simple form of functional relationship. More interesting forms relate two or more characteristics. Indeed, not only are they more interesting but they open up the door to more mechanisms for preventing data errors. A simple example of such a functional relationship would be that between the number of sons in a family and the number of children.

Obviously the former cannot exceed the latter. More generally, the sum of the cells of a table should equal the marginal. For example, if we categorize the employees in a corporation into age groups: <20, 20–29, 30–39, ..., then the sum of the employees in the different age groups should add up to the total number in the corporation.

Often these sorts of relationships are called 'rules', because they describe the way the data should behave. Sometimes they can be quite elaborate, and involve multiple characteristics: the three-year-old girl with two children, for example. Note that neither of these characteristics is suspicious, in and of itself. Three-year-old girls exist. People with two children exist. But put them together and we have an undoubted error in the data.

It must be obvious from this that an unlimited number of rules could be constructed. We could write down rules for two-year-old girls with one child, four-year-old girls with five children, four-year-old boys with a child, and so on. In an attempt to overcome this, researchers have developed systems which find 'covering rules': a rule for children of any sex, five years old or less, with more than 0 children would cover quite a few special cases, for example. Nonetheless, rule systems can become quite complicated.

Functional relationships tell us about relationships that must hold (number of sons plus number of daughters equals number of children). Statistical relationships tell us how we would expect different characteristics to be related, even if the relationship is not perfect. Again, they can apply to individual characteristics, but more interesting ones apply to multiple characteristics. While it is possible that, in a data set describing a large population, there may be a few women with more than twenty natural children, it is very unlikely that there would be many of them. A large count of such people should arouse suspicion: while it is not definitive proof of an error in the data, it indicates that closer investigation is merited.

A family with twenty children is an example of an *outlier*, something we briefly encountered in chapter 5. Most families have far fewer children. In general, anomalously large or small values are outliers. We heard of one example involving a surprising abundance of doctors apparently born on 11 November 1911 in a computer database. While it is entirely possible that some might have been born on that date, it seemed odd that so many should have been. One could imagine a recruitment boost increasing the number of medical students enrolled about twenty years after they were born, so it is just within the bounds of reason that there might be an unusual number of doctors born in a certain year, say, but on a particular day? Surely that is requiring too much of the imagination – and a more likely explanation is that there is an error in the data. Indeed, in this case the cause will be one with which modern users of computers and the web will be familiar. The age of some of the doctors was unknown, so this information could not be entered, but the data entry program insisted on an entry. An obvious null choice was 00/00/00, but the software developers had spotted this, and made it an illegal entry. The next most obvious choice was 11/11/11 – 11 November 1911.

This example involves only a single characteristic – date of birth. Other outliers only become apparent when we look at multiple variables simultaneously. We have already seen examples of this involving functional relations – a three-year-old girl with two children – but they can also involve statistical relationships. A ninety-year-old man is unusual, but certainly possible; a man who works as a professional athlete is not unknown; but put these together and again it perhaps justifies suspicion.

Other sophisticated statistical tools are based on theoretical or empirical laws for how data behave. An example of this is given by the *Benford distribution*, which has an interesting history. Before electronic calculators were developed, the easiest way to multiply large numbers together was to use tables of logarithms: adding logarithms is equivalent to multiplying numbers. While using such tables, Simon Newcomb (1835–1909) known particularly for his work in astronomy and economics, noticed that the earlier pages of his book of tables were much more worn than the later ones. It seemed that, in the sort of data he was studying at least, numbers beginning with 1 were more common than numbers beginning with 2, which were more common than numbers beginning with 3, etc. He went on to formulate a mathematical law relating the initial digit and their frequency of occurrence. Later, Frank Benford, after whom the law is named, noticed the same phenomenon, and tested it on several real data sets. This is curious and all very interesting, but how is it useful in detecting distorted data? Well, the law tells us what naturally occurring data (under certain conditions) should look like, but, and this is the important part, *it is not an obvious description*. It is not as obvious as, for example, the fact that the *last* digit of a set of numbers is often uniformly distributed – with the digits 0, 1, 2, 3, ... occurring about equally often. In fact, the very subtlety of the law means that it is very difficult to make up sets of numbers which appear realistic – for which the initial digits follow the Benford distribution. So, if people try to fake data, they generally produce data sets which do not follow the Benford distribution. This phenomenon has been used in detecting fraudulent tax returns and money laundering, as described in chapter 7.

In general, simple graphical tools can be highly effective in detecting anomalies in data. After all, the human eye and brain have evolved so that they can detect anomalies: a sudden movement in the grass may tell you that a tiger is lurking there, about to pounce. Figures 4.1 and 5.1 illustrated this human facility. In figure 4.1, the anomalous peaks at certain values are obvious. In figure 5.1, the change in the pattern of customers at a certain date is very readily apparent. These things would be vastly harder to see in simple tables of numbers.

Graphical displays have also been constructed with the explicit aim of detecting anomalies. In figure 6.4, each point shows where a missing values occurs in a collection of 1012 elderly women (the rows of the display, V1) measured on forty-five variables (the columns, V2), in a study aimed at constructing a screening instrument for osteoporosis. This graph immediately tells us which women (rows) have many missing values, and hence might be dropped from the analysis, and which variables have many missing values, and so do not contribute much information. The display also tells us that some variables are missing in pairs or triples (e.g. at the

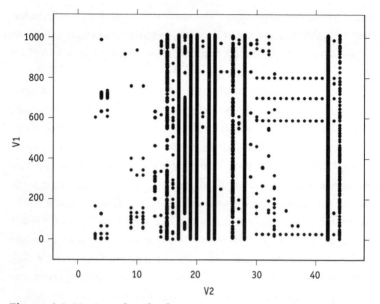

Figure 6.4 Missing value plot for responses to a questionnaire on osteoporosis.

lower left of the figure), which might be revealing about what is going on in the data.

Since it would be unwise to assume that any data set is perfect, an obvious question is how imperfect is it? How many errors are there in a data set? Estimates of this can be obtained in various ways. One can, for example, plot the rate of detection of errors against time. If, in the first week, 1000 errors are found, in the second week 200, in the third week 50 and so on, one could extrapolate to estimate how many would be discovered if one continued the exercise, and how many would remain. Of course, this will be only an estimate (and there is always the possibility of discovering a major, previously unrecognized, type of error which pervades the data).

This example also illustrates that a sort of Pareto law or 80/20 principle works in detecting errors. One is likely to discover most of the errors early on, and it will become more and more difficult to detect errors as one proceeds. At some point, then, one may decide that it is not worth looking any more: one believes the data are 'ninety-eight per cent clean', so that it is not cost-effective to continue. The cost of detecting the remaining small number of errors would outweigh the loss of revenue (or whatever) of ignoring them. The same issue arises in detecting fraud in banking and other places: if the loss due to fraud is $10 m, but it would cost $20 m to prevent it, it is probably not worth looking.

A related method of estimating the number of errors is to see how many of the same errors are detected in searches by two independent people. If they have hardly any in common it suggests there are many out there still waiting to be found, but if they overlap almost completely, then it suggests they are both picking up most of the errors, so that few remain. Of course, such methods are based on various assumptions.

A more active approach involves planting deliberate errors and then seeing how many are discovered. This is a sort of database version of the capture–recapture method used to estimate the sizes of animal populations. In this, a sample of animals is caught, marked or tagged, and then released back into the wild. Then another sample is caught. From the proportion of the new sample which are marked or tagged, one can estimate the total size of the population.

There are also less subtle ways of detecting errors. I am afraid I have used these when seeking typographical errors in proofs of my books in the past. I have sometimes paid graduate students, with an additional payment for every typo they find. (I am relieved to report that it did not make any of them rich.)

Preventing errors in data

Detecting errors in data is all very well, but it would obviously be more attractive to prevent them from arising in the first place. Detection can be helpful in this regard, since it can lead to the identification of causes of errors, so that errors can be prevented in the future. Perhaps it is a particular measuring device or researcher which is causing most of the problems.

Strict data entry protocols are an effective way of preventing errors. For example, carefully designed data capture forms can eliminate many unnecessary errors. The pharmaceutical industry has a wealth of experience in this area, with the design of case record forms for capturing the details of patients as they progress through a clinical trial of a new medicine. As a very simple example, if each of a series of questions has the possible answers 'yes', 'no', and 'don't know', then it makes sense to put these in the same order. (There are exceptions to this. If the aim is to force people to think, as in a school test, it would probably be silly to put all correct answers in the same position!)

Pharmaceutical companies also often use a system of double-entry keying of data to reduce data entry errors. In this approach, the case record forms which contain the written data are entered into the database twice, by two completely independent operators. In general, the probability that the same mistake will be made by both of these is very small. Values on which they disagree can be checked to see which was correct.

A similar idea is used with bankers' cheques, where the value of the cheque has to be written both in numerals and in words. It is very unlikely that both would contain the same error – although, of course, this cannot prevent the account holder from deciding to write a cheque for the wrong value. That is an error at a different level.

Of course, more and more data is now entered directly into a computer, and this opens up all sorts of additional possibilities. In particular, the idea of *defensive data checkers* is an important one. Here, the program monitors the data as it is entered, using the error detection methods described above. We have already seen an example of this in the program which would not allow 00/00/00 as a birth date. Unfortunately, as we saw, this program was not quite clever enough, and it did allow 11/11/11. These sorts of ideas sometimes go so far as to require the entry of redundant data: if respondents in a study are asked to enter the number of sons they have, the number of daughters they have, and also the number of children they have, then the computer can check for consistency. In general, the functional and statistical relationships discussed in the previous section can be used to flag contradictions automatically.

Sometimes people deliberately distort their answers because the questions are socially sensitive or personally delicate. Tools such as the randomized response method, mentioned in chapter 5, can help to sidestep these sorts of problems. In general, careful study design is important in reducing the rate of errors. Pilot studies, trying out data capture tools and methods, are vital in this context. In a similar sort of vein, sometimes the statistical power arising from repeating measurements, discussed in chapter 5, can be used. By averaging repeated measurements of an object, the overall error in the average value can be reduced as much as we want (or can afford).

In general, however, no matter how hard we try, we cannot prevent the unexpected from corrupting our data. Michael Berry and Gordon Linoff (2000) give a beautiful example in their book on data mining, quoting a company executive as saying that 'The data is clean because it is automatically generated – no human ever touches it.' But they then remark that twenty per cent of the transactions in the data set described files that had apparently arrived before they were sent, due to clocks not being set on computers.

Before I leave the area of preventing data errors, there is one other topic I should mention. This is the possibility of hiding errors. For example, suppose I suspect that ages have been entered incorrectly: I might suspect, for example, that there has been a tendency to round ages down (e.g. perhaps people aged thirty or thirty-one have a tendency to report their age as twenty-nine, those aged forty or forty-one to report their age as thirty-nine, and so on – a suggestion obviously linked to the data heaping phenomenon described above). If I suspect this has taken place, I can try to alleviate any problems which arise by grouping the data into age groups: into those aged twenty-eight to thirty-two inclusive, thirty-three to thirty-seven inclusive, thirty-eight to forty-two inclusive and so on. If distortions of the kind I expected have taken place, they will not lead to distortions in my grouped figures. On the other hand, I have clearly sacrificed some accuracy by doing this.

Living with errors in data

So suppose we suspect that we have failed to prevent some bad data leaking through (perhaps we can see the horrible gaps in the data we do have, showing that there are missing values). Suppose that we have run our data checking tools over the data and they have thrown up some apparent errors: we see, from the data, that apparently 450 of the people living in a village of size 300 are college educated, or we see that the rainfall over the past year was apparently such as to cover the entire country to the depth of 30 m. It is obvious that something is wrong. Either the 450 is too large, or the 300 is too small, or something else is wrong; and surely that 30 m cannot be correct. What do we do now?

Occasionally it is possible to go back to the source of the data to check. This is one powerful argument for checking the data as early as possible. But if one is conducting a data mining exercise a year after the data were collected, and only at this point does one find errors in the data, then it may well be impossible to go back to those who originally collected and recorded the data to ask exactly what happened. In general, sometimes the data were collected long ago, and those who collected it have long since moved onto other things, and can no longer remember, even if you were successful in tracking them down.

Based on our experience of such things in the past, we might be able to conjecture a reason for the errors: a misplaced decimal point, or something similar, and we might be sufficiently confident to 'correct' it. On the other hand, we might have no idea of what caused the error, or not be very confident of our explanation. What then can we do?

One possible strategy is to model the data distortion process, so that all subsequent analyses and conclusions include this model. Let us take missing values as an example.

In this case, we may well know that the data have problems – after all, we can see that there are great gaping holes in the data where there should be numbers. Now, missing values can arise for various reasons. In some situations, some of these reasons will lead to data being missing completely randomly. For example, perhaps hospital outpatients miss appointments because of public transport problems, and maybe these occur unpredictably, and at random. In this case, the data we have observed is a random sample from the complete data which we wanted to observe. This means that, in principle at least (the practice may be rather different), it is straightforward to analyse the data to arrive at conclusions: we just regard it as a random sample from the complete data set. Our 'model' for the data distortion process is that the fact that a data item is missing tells us nothing about the values it would have had, so we can ignore the fact that it is missing.

In fact, sometimes it is rather difficult to analyse incomplete data sets because the computer programs assume complete data. For example, they might require entries for all of the items in a questionnaire (think of the example of the birth dates of doctors, mentioned above). If one believes that any missing data are randomly missing, then one could simply drop incomplete records from the database. Although this will result in a smaller data set, it will not distort its overall statistical properties. Of course, if many records have some missing values, this approach risks dramatically reducing the data set size. I have already commented that it is not uncommon for *all* records to have at least one missing value. In such a situation, adopting the approach of dropping incomplete cases would mean that one was left with no data at all to analyse, so this is certainly not a universal solution.

An entirely different, and altogether tougher, situation arises when a data item is missing *because of the value it would have had if it had been observed*. This is a rather awkward concept, so let's have an example. Perhaps those hospital outpatients who are more seriously ill are more likely to miss their appointments simply because they find it difficult to get to them. Now it would be extremely dangerous to base our analysis, and to draw conclusions, just on the data we had observed, treating it as a random sample from the complete population. If we did this, our conclusions are likely to be very biased: we might conclude, for example, that the treatment was working, because we only saw the patients who were not seriously ill. In the worst case scenario, perhaps all the patients receiving a particular treatment get worse, but those who were initially most severely ill died. If these were regarded as missing in the analysis, it could mean that the average severity of those still alive at the end of the trial was less than the average severity of all the patients at the start, simply because the sicker ones had dropped out. Our conclusion would be the very opposite of the truth.

This is a sample selectivity issue, a type of problem that we have already encountered. We can try to tackle it by using what we know (or conjecture) about the cause of the missing values. If we believe that severity of illness is likely to be correlated with probability of dropping out, we can build a statistical model for this. This model then allows us to adjust the predictions based on the data we *have* observed. The result is a model applicable to the entire population.

You will probably suspect that this simple explanation conceals a great deal of complexity, and you would be right. However, this basic idea underlies the approaches developed by the economist James Heckman, which led to him being awarded the Nobel Prize for Economics in year 2000. It has to be said, however, that such models are very sensitive to the assumptions one makes about why the unobserved data are missing. If your assumptions are wrong, your conclusions may well be. The fact is that you can't get something for nothing. It is another example of the familiar Free Lunch Theorem, which simply says 'there's no such thing as a free lunch'. If the data are missing, you have to get the missing information from some-where, and assumptions or ideas about how and why they are missing is one possibility.

Sometimes another approach is possible. This is to build models that are not very sensitive to the values the missing items might have taken. Such robust models will tend to lead to similar conclusions whatever the complete data would have looked like. Of course, this also is a delicate issue. The most robust model of all would simply ignore the data. For example, if I estimate the average age of drivers in the US to be thirty-five, without taking into account any data at all, then it would not matter if some was missing: I would always conclude 'thirty-five'. On the other hand, if I estimated the average age based on a survey carried out late at night, I might under-estimate the true average age because of a preponderance of younger drivers at that time. As with all data analysis, intelligence, expertise, experience and above all common sense is required to arrive at valid conclusions.

Relevance, timeliness and beyond

I remarked in the opening section of this chapter that issues in addition to accuracy were important when assessing the quality of data. These issues include relevance, timeliness, accessibility, clarity and consistency – and you may well be able to think of others.

Relevance is closely bound in with the issue of measuring the right thing. A relevant measurement measures its intended target. It is no good having very accurate measures of things which do not capture what one wants. League tables provide a good example. A ranking of universities according to the quality of their research output may be of limited value to a potential student who is more concerned with the quality of the lectures and how much tutorial support she is going to get.

On the other hand, producing appropriate measures is often very difficult. Sometimes it is necessary merely to do the best one can, acknowledging that it needs to be improved. Often *proxy variables* are used in place of what one really wants to measure. These are variables which, one believes, behave in a way similar to the thing one is really interested in. For example, we could measure extent of hunger by measuring how quickly food is eaten, how much is eaten, how many obstacles will be overcome to reach food, how long it has been since food was last eaten, and so on. None of these are really what one wants, but, depending on the use to which the measurement is intended to be put, any of them may be perfectly acceptable as a proxy measurement of hunger. A more familiar example arises with the breathalyzer. Here we really want to measure the extent to which someone is intoxicated and is therefore unsafe to drive, but, as a much more easily measured proxy, we instead measure the amount of alcohol in their breath. This, of course, often serves as a precursor to another proxy variable, albeit a more accurate one – the blood alcohol level. Yet another familiar proxy variable is weight loss when dieting. People generally don't really care what they weigh – what they are really interested in is what they look like. As Arnold Schwarzenegger put it in his bodybuilding days: 'Look in the mirror, not at the scales.' But weight is easily measured and is, at a broad level at least, related to observable fat on the body. I say 'at a broad level' because other things also affect weight. Body fluid, in particular, has a big effect. Some rather dubious weight loss regimes capitalize on this: stop drinking for a day and you will lose several pounds. Of course, you won't have lost any fat, and it won't have done you any good, but the proxy measure of weight appears to show an improvement.

Data are *timely* if they are there when they are needed. Data on which of a supermarket's shelves are running dangerously low are of limited use if they are not available in time for the staff to restock the shelves before the shop opens tomorrow. Note that sometimes different measures of data quality work in opposite directions. Timeliness and accuracy provide an example of such a pair. Economic time series, for example the rate of inflation, the Gross Domestic Product, and so on are needed as soon as possible, so that the government can monitor how the economy is performing and take appropriate action if necessary. However, producing such figures is a far from straightforward task. It involves collecting data from a wide variety of sources and then condensing all this data down using sophisticated procedures, but if all this takes too long, it will be too late for the government to take effective action. The consequence of this is that often the initial figures are later adjusted. Inevitably this sometimes causes difficulties – when one finds out that the inflation rate wasn't too bad after all, and one needn't have rushed into action, for example.

Data must also be *accessible*. This means not only that computer files containing the data exist and can be downloaded, but also that the data must be readable without extensive processing. Early in my career I experienced the sharp end of this, when I was faced with analysing some satellite image data. The data files were sent, with no problem (though, it being early days, they were on magnetic tape), but it then took me six months of preprocessing before I could actually read the data prior to beginning any analysis.

Connected with the issue of accessibility is that of *clarity*. An important part of this is that the data must be accompanied by a description of the data (sometimes called a data dictionary or, as we have seen, the *metadata*). A table of numbers, no matter how accurate, timely and easy to read it is, will be useless if you do not know what the numbers actually represent.

The data must also be consistent if they are to be useful. It will be of limited value if different parts of the data contain contradictions.

All of these attractive properties of high quality data – and any others that you can think of – do not come for nothing. *Quality costs*, as we might say, and we have already seen this in the contexts of preventing and detecting errors in data. At some point we need to consider whether we can live with a lower quality of data (using a cheaper but less accurate proxy measurement, taking a day longer to get the data, etc.) or if it is worth the expense of improving it.

The morality of good data

This chapter has been primarily concerned with bad data: data that fail to properly represent the phenomenon they are supposed to be describing. Bad data, as we have seen, are likely to lead to mistaken conclusions. Good data, by contrast, can lead to correct conclusions. That they need not do so is because data, in themselves, do not tell us anything: data have to be analysed, have to have their meaning squeezed from them, in order to tell us things, and it is entirely possible that mistakes may be made in the analysis.

However, since good data do permit the possibility of valid conclusions (whether or not one likes those conclusions is a different matter), such data often carry an aura of moral rectitude. Thus, extreme accuracy is often seen as a positive, upright thing, implying care and attention to detail, and suggesting that someone is painstaking and cannot have the wool pulled over their eyes. The impersonality of accounting is another aspect of this: the numbers are regarded as objective, and hence representing truth. In this sort of vein, M. Norton Wise (1995) has remarked that 'precision instruments are packaged trust. And the trust they carry is a social accomplishment'. If you want me to trust your data, you have to convince me that they are accurately measured. If you can convince me of that, then I will believe what you say – about the data and beyond.

On the other hand, in some situations it can be useful to preserve some ambiguity: that way one does not have to commit oneself. For example, in the context of health data, Ian McDowell and Claire Newell (1996) have remarked that 'there is a historical tension in approaches to health between those who prefer to keep the concept somewhat imprecise, so that it can be reformulated to reflect changing social circumstances, and others who define it in operational terms, which often means losing subtle shades of meaning'. Evolution, of course, hinges on this sort of ambiguity. If all animals of a certain species were identical, then they would all be equally vulnerable to dangers such as diseases. In fact, however, they are all very slightly different, so that they have different chances of surviving.

This merit of uncertainty or ambiguity in measurement is related to the idea that sometimes one wants to avoid analysing data in too much detail. While a sophisticated statistical analysis might yield half of a percentage point of additional accuracy in predicting future values, one needs to bear in mind the possibility that the future may not be exactly like the past. It may be that future data arises from some sort of process that differs from past data, at least to the extent that that half a per cent becomes irrelevant.

Tempting though it is to finish this chapter on an upbeat note, it is probably more honest to leave it on a note of caution. Data, as we have seen, are generally not 100 per cent correct, and can have all sorts of problems. A good working position, as far as data quality goes, is therefore one of suspicion. If a data analyst is presented with a data set which appears to be error free, to have no missing values, no strangely large or small numbers, and so on, then he or she might legitimately ask if it has been preprocessed in some way. Have artificial substitutes been created to put in place of missing values? Or have incomplete records simply been dropped from the data? Have excessively large numbers been reduced? All of these things, while possibly perfectly legitimate, have moved the data away from the mechanism which generated it. If the data cleaning has been informal and subjective, it is possible that all sorts of biases might have been introduced. One has no way of knowing. One's descriptive conclusions are descriptions of the entire process that led to the recorded data. Apparent perfection in a data set should motivate caution: you may not be being told all there is to know.

CHAPTER SEVEN

Deception and
Dishonesty with Data

Anyone can easily misuse good data.
William Edwards Deming

Overlooking, concealing and inventing reality

The Challenger space shuttle example in chapter 6 shows what can happen if some data are missing. Of course, there is no doubt that that was a genuine oversight, a mistake about how to analyse data and, in particular, about how to handle missing data. Sometimes, however, it might not be a mistake. Take surgeons, for example. An easy and obvious measure of their performance would be the proportion of their patients for whom the surgery leads to full recovery. At face value, one would prefer to be operated on by a surgeon who has a ninety per cent success rate, rather than one with only a fifty per cent success rate. But this is based on the assumption that the two surgeons are operating on similar populations of patients. If one subsequently learns that the first surgeon turns away any seriously ill patient (so the outcome of their operation on such patients is unknown), and operates only on those whose illness is very mild, then one might take a different view. Indeed, if we take this to the extreme of a surgeon who will only operate on patients he thinks will recover naturally anyway, then the fact that ninety per cent is not 100 per cent gives an altogether more sinister impression. A fifty per cent success rate for a surgeon who willingly takes on the toughest of cases may be much more impressive than the other's ninety per cent rate.

This principle applies in other areas. Schools are often ranked in league tables, in terms of such things as their performance in national examinations and tests. But one's real interest is probably the extent to which the school *adds value* to their intake of children. A low average score from a school in a deprived inner city area may be much more impressive than a far higher score from a pampered middle class or wealthy area. The question is how much the school has contributed to their students' education, advancement and progress, and not whether the schools have coasted while the students do what they would have done anyway, regardless of the school's contribution or lack of it.

Perhaps an extreme of this sort of thing is the anecdotal report of someone's grandmother, 'who smoked two packs of cigarettes a day all her life, and lived to be 95', so 'proving' that smoking can't really be bad for you. Selecting just the data which are conveniently to hand (that about one's grandmother) and ignoring all the other data (all those who died from smoking-related diseases) can be very misleading.

Overlooking data, as in the Shuttle example, is one thing, but sometimes far worse happens. After all, it is entirely possible that the data you do collect might well disagree with your preconceived ideas or prejudices. It might contradict the ideas you have long held dear, or on which you have based your reputation. So, if collecting data is an expensive process and might not help you prove your point, surely life would be far easier if you simply made the data up! That way it won't cost anything and you can be sure it will support your argument: you can be sure it will prove what you want to prove. As someone once said to me at a conference on performance indicators: 'It is often easier to fiddle the figures to meet the targets, than really to improve performance.'

Of course, this is not a new problem. Harold Cox, quoted in Sir Josiah Stamp's (1880–1941) 1929 book *Some Economic Factors in Modern Life*, cites a judge as saying 'The Government are very keen on amassing statistics – they collect them, and they raise them to the nth power, take the cube root and prepare wonderful diagrams. But what you must never forget is that every one of those figures comes in the first instance from the [village watchman], who just puts down what he damn pleases.' One study by the US Census Bureau suggested that between three and five per cent of census enumerators made up some of their data. Indeed, this practice is so common that it has been given the colourful name of *curbstoning*: instead of bothering to visit the household, the enumerator simply sits on the curbstone and makes up the numbers.

Investigators making up the data is one thing: different again is when the original source makes it up. In social surveys on sensitive topics, for example, respondents might be under heavy social pressure to lie. Thus, surveys on sexual behaviour, smoking, drinking and even number of hours spent watching television might well lead to consciously distorted results. Tax returns and expenses claims are also susceptible to conscious distortion at the level of the individual originator of the data. The aim, as above, is to deliberately distort things so as to mislead.

One might think of such cases as being spread out along a continuum of moral turpitude or social acceptability. Reporting fewer than the true number of hours spent in front of the TV is probably fairly minor compared with hiding a few million dollars in one's tax return. We have all also heard of minor distortions of the truth, or even major alterations, aimed at boosting someone's reputation. Regrettably, this has even sometimes occurred in science. Sir Francis Galton (1822–1911) made a cautionary remark about such things in his book *Meteorographica*, published in 1863: 'Exercising the right of occasional suppression and slight modification, it is truly absurd to see how plastic a limited number of observations become, in the hands of men with preconceived ideas.'

Take the case of Louis Pasteur (1822–1895). Pasteur was undoubtedly a great man. He discovered that most infectious diseases are caused by tiny micro-organisms – now called germs – and this understanding has become a linchpin of modern medicine. The advance led to better hospital practices to control the spread of disease, to methods of vaccination based on using weaker forms of a microbe, to the discovery of even smaller disease-causing agents, now called viruses, and to the development of heat treatment methods (now eponymously called 'pasteurization') to kill the microbes in food so that perishable foodstuffs could be stored without rotting. There is no doubt that Pasteur made tremendous advances which have benefited humanity immensely. No wonder, then, that he is generally regarded as having saintly characteristics. The surgeon Stephen Paget called Pasteur 'the most perfect man who has ever entered the Kingdom of Science' and also said 'here was a life, within the limits of humanity, well-nigh perfect ... here was a man whose spiritual life was no less admirable than his scientific life'.

A closer look at the man and his practices shows that perhaps he was not as pure as has been described. Some of his central work hinged on demonstrating that micro-organisms did not spontaneously generate, but were carried from elsewhere – in the air for example. This conflicted with the ideas promoted by Felix Pouchet, who had accumulated evidence favouring spontaneous generation. Pasteur replicated some of Pouchet's experiments, and carried out others, under very rigorously controlled conditions, claiming to approach them without preconceptions, but his notebooks show that he regarded as 'successful' those experiments tending to refute spontaneous generation, and as 'unsuccessful' those conflicting with his preconceptions. Put another way, he was choosing from his observations those data which tended to support his viewpoint. He was deliberately biasing things in favour of his preconceptions.

Another example of how data can be distorted is given by the Nobel Prize-winning American physicist Robert Millikan (1868–1953). Millikan wanted to know whether electricity was a continuous fluid-like substance, or was composed of myriads of tiny identical particles of electricity. The notion that electricity might be particulate is central to the atomic theory of matter. It is a building block of twentieth and twenty-first century science and technology. If electricity is particulate, then any difference between electrical charges must be a discrete multiple of a smallest value – the charge on a single particle, the electron. Based on this idea, Millikan measured some very small charges, to see if they were multiples of some smallest value.

The basic principle behind Millikan's experiment was to hold tiny charged droplets of water (and later oil) in an electric field, balancing them against gravity. When the field was switched off, the particles would fall, and by observing their rate of fall, and taking into account the viscosity of the gas through which they were falling, he was able to calculate the mass of the particles. This in turn enabled him to calculate the force exerted by the electric field, and hence the charge in each particle. As he later wrote: 'Charges actually always came out easily within the limits of error of my stopwatch measurements, 1, 2, 3, 4, or some other exact multiple of the smallest charge on a droplet that I ever obtained.'

Millikan published his results in the *Philosophical Magazine*, including all of his results, but indicating, with a system of stars, how accurate he thought each result was. Weighting his calculations according to this subjective measure of quality led to his estimate of the charge on an electron.

Of course, a less scrupulous researcher would simply have dropped the 'lower quality' results, and no-one would have been any the wiser. By including them in this way, Millikan was being very honest, and explicitly admitting to the fact that he was prepared to adjust his results. This, of course, prompts the question about whether his scientific objectivity could be relied on. Worse, in the paper, Millikan said of some of the lower quality results: 'I would have discarded them had they not agreed with the results of the other observations.' Is that not an explicit admission that he was prepared to select his data to support his preconceptions?

For Millikan, history, and huge amounts of further research using entirely different theoretical approaches, has lent support. But couldn't things have gone the other way? What if he had believed that electricity was non-particulate, and had selected only those experimental results favouring that hypothesis? Where then would be his Nobel Prize?

Of course, it is not unusual for scientists to select just some of their experimental results to use and to include in a subsequent analysis. After all, if you accidentally bumped a dial on the apparatus, sending one of the control parameters through the roof, you would feel entirely justified in omitted the resulting reading from the analysis. Does this selectivity mean you are being dishonest? What if one of the readings was so far out of line with the others that you doubted that it was correct, even if you could not explain exactly what had gone wrong? Obviously we can extend this line of argument, and at some point it is not clear whether or not exclusion of readings is justified. A general principle might be that one must have good reasons for such exclusions, reasons which others can assess and hopefully accept. Millikan's initial results, about which he was so disarmingly and dangerously honest, were described in his first paper, published in 1910. But he learnt from the experience. In another paper, published in *The Physical Review* in 1913 describing similar experiments but under much improved conditions, he wrote 'Table XX contains a complete summary of the results obtained on all of the 58 different drops upon which complete series of observations like the above were made during a period of 60 consecutive days,' but his notebooks contain jottings suggesting that this might in fact not be the complete truth.

Rejecting readings may be a rational strategy when one has sound reasons for believing them to be unreliable, but one should be wary of doing so when the reasons are related to the theory one is testing. The issue is the familiar one of sample selectivity. In general, while one may have sound reasons for rejecting some observations, there is always a risk in doing so. Perhaps your reasons are wrong, and the observations really are perfectly good. Or perhaps they are unusual for some alternative and unsuspected reason. A case of this is described by István Hargittai (2002):

> When Herbert Brown's co-worker tested 57 substances for a certain reaction and 56 of them showed consistent behaviour, the co-worker suggested dropping the exception and publishing the results from the other 56. The 56 substances that gave the expected reaction were standard materials in Brown's lab. They had been tested and purified by a careful procedure. The odd 57th compound may have contained impurities and it would have been easy to write

it off and go ahead with their usual procedure. However, Brown insisted on repeating the experiment. It was in his nature not to rest until he got to the bottom of things. As it turned out, the 57th compound behaved in an understandable way when they let it react for a more extended period of time than the rest. It contained a double bond for which it took a longer time to add a B-H bond, which it did, and this recognition led to the discovery of a new chemical reaction, hydroboration, and to Brown's Nobel Prize.

Controversies about the selection of data, either consciously or sub-consciously, arise occasionally in medical research. Repeated studies have now clearly demonstrated that there is no link between the triple MMR vaccine and autism, but early controversial work, apparently based on distorted samples, suggested that there might be a link. As a consequence of the concern raised by this mistake, the overall immu-nization rate in the UK fell to below eighty per cent – substantially below the ninety-five per cent required to ensure that epidemics cannot spread in schools – and children fell ill. Another medical example is given by the Debendox case. This was a drug for preventing severe nausea and vomiting during pregnancy. Since these symptoms are so common, the drug was widely used – about thirty per cent of pregnant women in Australia took it in the late 1960s. But birth defects are also quite common: about five per cent of Australian births have a significant problem of this kind. From observing the two things in conjunction, it is not a great leap to conjecture that Debendox causes defects. The Australian obstetrician, William McBride, did indeed leap to this con-clusion and, partly because of his promotion of this viewpoint, Debendox was withdrawn from the market. This is despite the fact that very extensive later studies have established that there is no link. As a consequence, no drug to ease those symptoms during pregnancy was marketed, so that thousands of women suffered unnecessarily.

Of course, the fact is that Pasteur and Millikan were right. Their initial ideas, supported by the perhaps shaky evidence they first pro-duced (but painstakingly, and with immense difficulty, insight and determination), have subsequently been tested, checked, examined under different conditions, and the predictions arising from their ideas have been explored in vast depth. Perhaps their ideas have been modified and adapted as we have learnt more about the world around us, but their ideas are still there at the core. Unlike history, science is not written by the victor, but rather the ultimate victor is he or she who is vindicated by later experiments. Like history, when this happens, it is only natural that retrospective descriptions do not dwell on the shortcomings or misdirections of those who caught what ultimately turned out to be a glimpse of the truth.

The work of Sigmund Freud has come under a vast amount of criticism: of the non-refutable, and hence non-scientific nature of his claims, of the lack of any remotely good match between his theoretical descriptions of how the mind works and of how the evidence (the data) shows it to work, of the doubt that the 'method' of psychoanalysis has any beneficial effect whatsoever, and so on. Freud said that support for his theories was provided by eighteen cases which he had successfully treated, though his descriptions of what they were cases of, and of the experiences they had suffered, changed over time. Unfortunately, the situation is even worse: it seems that these cases did not exist at all. From correspondence which was only published in full in 1985, it is apparent that Freud had treated thirteen such cases, without any successes. Freud invented his data.

Sometimes people who make up their data seem to get carried away. Perhaps because they are not detected the first time, they think they can do it again and again, but this cannot go on for ever. The case of William Summerlin is a classic.

Summerlin was an expert in skin transplants. Such transplants are normally rejected, unless the donor has an identical genetic make-up to the recipient. However, Summerlin described how rejection could be prevented by a special treatment of the grafted skin, and demonstrated this with skin and other grafts in small mammals. Unfortunately, others could not replicate his results (an example of the self-correcting nature of science in progress?). Summerlin's boss at the Sloan Kettering Institute in New York, Robert Good, decided that they should write a paper about this issue. Summerlin went to see him, taking with him eighteen white mice which had had patches of black skin transplanted, to demonstrate that his methods had worked. Unfortunately, within a few weeks, a laboratory assistant noticed that the black patches on some of the mice looked odd. A little experimentation revealed that they could be washed off with alcohol. The different colours were not due to transplants from different coloured mice at all, but were simply the result of using a black marker pen. Once this had been discovered, the whole deceit rapidly collapsed. Close examination showed contradictions in Summerlin's reports. Data were missing. In a paper submitted to the *Journal of Experimental Medicine*, the numbers did not add up correctly. Some of the data were plain wrong.

Another classic case is that of John Darsee, a specialist in heart attacks who worked at the Harvard Medical School. Once again, some of his co-workers became suspicious, and reported their suspicions to the head of his laboratory, Robert Kloner. Kloner asked Darsee for his data, so he could check the results. But Darsee didn't have the data. Instead, he started creating some, taking measurements from a dog. He recorded these as Day 1, Day 2, etc., as if they had been taken at different days, but didn't bother to hide what he was doing from others in the laboratory. Is this because he had got away with it before, and thought he was invulnerable? Afterwards he claimed that he had discarded the original data, but the suspicion must be that he had never had it, and that he had simply made up the results in his paper, not even bothering to fabricate data from which the 'results' could be derived. Further detailed investigation showed that this was not the only work for which he had invented things. Dozens of his papers, published while he was working at the Harvard Medical School and, previously, at Emory University, were shown to be based on imaginary data.

Although science is self-correcting, in some cases that correction can be a long time in coming. Perhaps especially in medical contexts, once a treatment has been established as effective there is a natural resistance to testing it by comparing it with others that are thought to be less effective. Indeed, such a comparison might be regarded as unethical, since one would be giving some patients a treatment regarded as less effective. However, if the 'effective' treatment is based on fabricated data, untold numbers of patients may suffer. Making up data is not a victimless crime.

Most of the examples above involved either overlooking important parts of the data or simply making up the data. We have seen that there is a gradual transition from, at the one end, perfectly reasonable mistakes or decisions about which data items to use, to, at the other end, deliberate and dishonest distortion of the facts. Sometimes, however, distortion occurs at an even later stage: the data, perfectly fine in themselves, might be misused, even deliberately so.

The public perception of statistics is fairly cynical in this regard, perhaps having come to *expect* statistics to mislead. Think of the quotation 'There are three kinds of lies: lies, damned lies, and statistics' (frequently attributed to Benjamin Disraeli, but probably due to Leonard Henry Courtney in an article published in *The National Review* in 1895), or of the quotation 'Facts are stubborn things, but statistics are more pliable.' Whenever the definition of unemployment is changed, people often assume this is to make the current administration look better relative to its predecessor.

Advertisers are not immune from a tendency to mislead by misusing data. We have already seen how some medical conditions, such as the common cold, in time clear up by themselves. Glossing over this essential fact allows one to imply that a cold remedy is effective.

In 1994, American Baseball players went on strike. The issues were complicated, but examination of the distribution of salaries shows how data can be manipulated. The average annual salary of players was $1.2 million. 'Average' here is calculated as the *arithmetic mean* which we met in chapter 5, by adding up the salaries of the 746 players and dividing by 746. To me, at least, a salary of $1.2m sounds perfectly respectable – and the owners of the club could certainly argue that they could not afford to increase this average. On the other hand, the *median* salary is only $0.5m. As we also saw in chapter 5, this is the salary such that half the players earn more than it, and half earn less than it. The players, using this notion of average, could argue that such stars as themselves certainly deserve more. Who then is right: the owners, basing their argument on the arithmetic mean, or the players, basing theirs on the median? Both the arithmetic mean and the median are averages, in the sense that they are representative values summarizing the distribution of salaries. But clearly they tell very different stories.

The reason for the difference hinges on how the salaries are spread out, in particular on the skewness of the distributions: there are many lower salaries (indeed, half of them are below $0.5m), but they become fewer and fewer the higher you get. So seventy-five players earn between $3m and $4m, twenty-six players earn between $4m and $5m, ten players earn between $5m and $6m. And only two players earn over $6m each year. The arithmetic mean balances many 'small' salaries with a handful of very large ones, giving something in between. In effect, the excessively large salaries pull the average up, making it much greater than the median. Which one you choose, mean or median, depends on what you want to find out, on what question you want to answer, or, one might say, on what story you want to tell.

Of course, it requires even more manipulation of the data to be able to say, as did an Australian Labour Minister, 'We look forward to the day when *everyone* will receive more than the average wage' (my italics). In fact, I have not yet discovered the average which allows one to do this – but then perhaps politicians have their own way with data.

It is said that a picture is worth a thousand words, and graphs are certainly an ideal medium for communicating numerical results – for conveying data. On the other hand, they are also an ideal medium for misleading people about numerical results. Simple malpractices can deceive the uninitiated. For example, by adjusting the range of axes of graphs one can give a completely distorted impression of the truth. Figure 7.1 shows how sales for a fictitious company have changed over time. Apart from the dip in the fourth year, things have progressively increased. It's looking good. But a closer examination shows that the scale of the vertical axis starts at ninety. If the data are replotted, with the full range of the scale, as in figure 7.2, a rather different picture emerges. Perhaps sales are just about

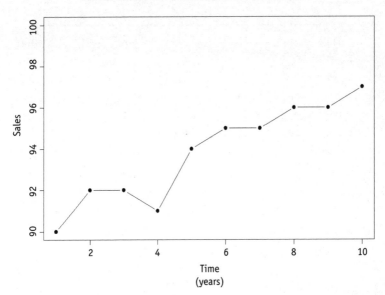

Figure 7.1 Sales plotted against time

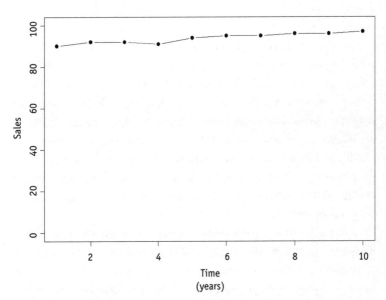

Figure 7.2 Sales against time, with the full range of vertical axis.

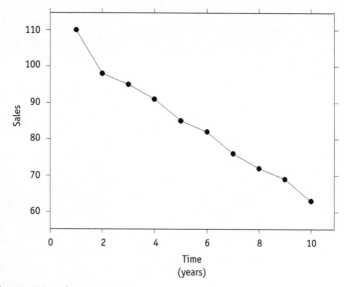

Figure 7.3 Sales over time.

remaining constant. In his classic 1954 book *How to Lie with Statistics*, Darrell Huff has a chapter devoted to what he calls 'the gee-whiz graph'.

Of course, there are circumstances under which the first plot would be the correct one (perhaps one wanted to draw attention to the dip in year four, which is barely noticeable in the second plot).

In fact things can be even more misleading. Figure 7.3 shows another fictitious plot of sales against time. Clearly, for this corporation, things are looking gloomy: sales have progressively declined over time.

Figure 7.4 puts an entirely different complexion on things. This plot shows cumulative sales. Although exactly the same information is conveyed by figures 7.3 and 7.4, a superficial examination can be very misleading.

The danger of being misled by inappropriate statistics is also illustrated by the death rates attributable to accidents in different kinds of transport. Table 7.1 shows figures from the European Transport Safety Council 1999 giving the number of deaths per 100 million person-kms for various modes of transport.

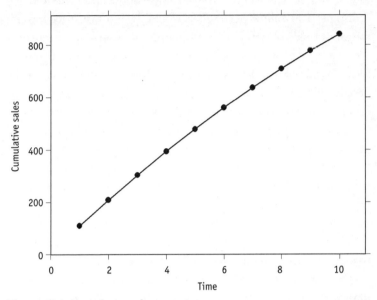

Figure 7.4 Cumulative sales over time.

Table 7.1 Number of deaths per 100 million person-kms by mode of transport (European Transport Safety Council, 1999).

Motorcycle/moped	16
Foot	7.5
Cycle	6.3
Car	0.8
Bus and coach	0.08
Air (public transport)	0.08
Rail	0.04

Likewise, figures for the US give the death rate for flying as 0.6 per billion miles and for road travel as 24 per billion miles.

It would seem from these numbers that travel by car is much more dangerous than travel by air. But is this right? The average plane journey is much longer than the average car journey. The US figures include local automobile travel, but perhaps travel on the rural

Interstate system is a more sensible comparison with air travel. Using deaths per 100 million persons km would suggest that space travel is the safest mode of transport of all, simply because all space journeys are so long. Perhaps it would be more relevant to give the figures in terms of number of deaths *per journey*. Moreover, the nature of the people killed in the two kinds of accidents differ. The road statistics, for example, include drunken young men who kill themselves.

In a similar vein, let us look at when road accidents happen. Road accident statistics for Scotland over the period 1994–1998 show that there were, on average, over three times as many deaths on the roads at 6 p.m. than there were at 6 a.m. From this, one might infer that it was over three times as dangerous to travel in the evening than in the morning, but of course, this fails to take into account the numbers of people travelling at those times. There are far more people on the roads in the early evening than in the early morning.

These transport examples are illustrations of a more general way of distorting data: compare figures relating to different things. We can compare the health of the population in two different cities by looking at the death rates – but if one city is a retirement haven while the other is a sports resort the rates may be very different. It might be that we want the raw comparison, but it also might be that we want to adjust for the age difference.

Distortion of data, deliberately choosing from data and manipulating data to one's own ends are not new phenomena. They are as old as data themselves. Lying is even possible without speech: animals have been known to deliberately mislead other animals. In the book *Reflections on the Decline of Science in England* (1830), which we have already encountered in chapter 1, Charles Babbage wrote: 'There are several species of impositions that have been practised in science, which are but little known, except to the initiated, and which it may perhaps be possible to render quite intelligible to ordinary understandings. These may be classed under the heads of hoaxing, forging, trimming, and cooking.' He described *hoaxing* as being intended to be ultimately revealed, to ridicule those who had fallen for it, but *forging* as being intended to last for a long time. He defined *trimming* as 'clipping off little bits here and there from those observations which differ most in excess from the mean, and in sticking them on to those which are too small', and *cooking* as selecting the data to support one's argument.

White lies: good reasons for concealing the truth

We saw in chapter 6 that many kinds of distortions can arise which damage and corrupt data. We have also seen in this chapter that sometimes those tasked with collecting data are not as rigorous as one might like, and that it has even been known for people to fabricate the data, or to bend it to support their aims. However, having said all that, sometimes there are ethically sound reasons for hiding the true data.

Perhaps a very simple example of this occurs in the double-blind clinical trial of new medical treatments, mentioned previously. One of the difficulties with testing new treatments is that people may react differently if they know which treatment they are receiving. In fact, as we have already discussed, people often show a response even when they are given an inactive placebo treatment. This sort of thing complicates experiments aimed at discovering how effective treatments are. One strategy for overcoming it was described in chapter 5 – to compare the responses of patients randomly allocated to either the active treatment or an inactive placebo in such a way that neither the patient nor the doctor administering the treatments know which each patient is receiving (perhaps the tablets are made up to look the same). A secret code, deciphered only *after* the data have been collected, shows which patient was allocated to which treatment. The point is that concealing the data while the experiment is being run is crucial to its success. In this case, hiding things is vital.

In order to run efficiently, governments need to collect data about their people. Many other bodies and governmental subsidiaries also need to make use of this data. However, businesses, people and other organizations will often only divulge data if they can be sure that it will remain confidential. This can pose a problem. Suppose, for example, that an academic researcher, studying the adequacy of the services provided by local government, receives detailed tables of information from a local government survey. Suppose these tables show that there is only one (unnamed) doctor earning a salary in a certain range, and who lives in a particular part of the town. It would then not take much initiative for the researcher to be able to discover the name of that doctor. This example applied to an individual, but similar situations could apply to schools, or corporations, for example. In such situations the data are no longer confidential. Not only has this broken any initial agreement to preserve confidentiality, but it also jeopardizes the future willingness of individuals or businesses to divulge data. A corporation will be unhappy about giving out information which a competitor could use to its advantage.

One way to tackle the problem is to restrict access to the data. This may require some complicated legal manoeuvres. An alternative is to adjust the data to try to make it impossible to discover things about individuals. Leon Willenborg and Ton de Waal, experts in such modifications of data so that it only reveals what you want it to reveal, define *disclosure control* as 'the discipline concerned with the modification of statistical data, containing confidential information about individual entities such as persons, households, businesses, etc. in order to prevent third parties working with these data to recognize individuals in the data and thereby disclose information about these individuals (Willenborg and de Waal, 2001).

Obviously such modifications require a great deal of care and delicacy. There would be little point in 'modifying' the data to the extent that it became useless for its intended purposes. Measures of *information loss* are used to quantify the extent to which a particular adjustment compromises the quality of the data.

Such modifications can be carried out in various ways. To take a very simple example, in the case of the single doctor above one could merge salary ranges, or merge regions of the town, so that no range had a single person in any one area of the town. Observed values can also be replaced by missing value indicators – as if the particular data item was not supplied (e.g. the doctor's salary). There is a subtlety here: often extreme values will be natural candidates for suppression in this way (e.g. the highest salary). The result could mean a significant distortion of the data (if all the highest values have been replaced by 'missing', then the average of those recorded will underestimate the true average). There is also the possibility of recognizing that something which has been coded as 'missing' means that it really had an extreme value. To sidestep this problem, a more advanced version uses a randomly imputed value, instead of a missing code. It means that the overall properties of the data are retained (if the imputation is done properly), but information on individuals is not divulged. In fact, this principle is sometimes taken to an extreme. Sometimes statistical methods are used to model the data, and then an entirely new artificial data set is randomly generated from this model. Again, it has the same statistical properties as the original data, but no confidential information on individuals is given out. Of course, a single new artificial data set could by itself be misleading. Typically, therefore, the exercise is repeated several times, so that users are provided with multiple samples drawn from the same distribution. Perhaps it is unnecessary to remark that such methods involve a sophisticated level of technical competence to do properly: the difficulty is that one cannot predict what users might want to do with the data, and so cannot be certain that the model fitted to the original data is adequate for its purpose.

In cases where the full data consists of an entire population (e.g. a census, or records of company employees), another strategy is to release only a subsample of the data. Care has to be exercised if different subsamples are released each time a request is made, since linking these samples can lead to identification of and discoveries about individuals.

The method of randomized response was described in chapter 5 as a tool for collecting data on sensitive topics. It enables accurate information on proportions in the population to be collected, while avoiding giving certainty about the answers any individual would have given. Similar ideas can be applied to process data before giving it out to other people. Another version of this same idea randomly swaps values between two records in the database.

Finally, yet another idea is to randomly perturb the data: to add a random error to the observed value so that, for example, someone whose true salary is in the $30–40,000 range has a small but non-zero chance of being recorded as lying in the $20–30,000 or $40–50,000 range.

Once again, it is obvious that these methods involve a high degree of technical sophistication.

Both the USA and the UK have Freedom of Information Acts. That of the USA was introduced in 1966 and became law on 4 July 1967. According to the United States Department of Justice Guide to the Freedom of Information Act, it 'firmly established an effective statutory right of public access to executive branch information in the federal government'. Likewise, the UK's Freedom of Information Act 2000, which came into full force on 1 January 2005, extended the right of access, by the public (individual or corporate) to information held by public authorities. In both cases, essentially every item of information must be made available on request (and on payment of an appropriate handling fee) unless it is specifically excluded from coverage.

Of course, the acts do not apply to private bodies. Indeed, it is in the interests of competition between such bodies that they can keep their commercial data confidential. For this reason, and because of the universal use of electronic means of transferring data, sophisticated data encryption schemes are used. These are rather different from the disclosure prevention schemes outlined above because they do enable the intended recipient, but only the *intended* recipient, to reconstruct the original data perfectly. Exactly the same sort of ideas are used, of course, in encoding credit card numbers and bank account details when they are transmitted across the internet. Military communications are likewise encoded, also for obvious reasons.

Forensic data analysis

This chapter has given many examples of how data has been made up or altered. However, it is very difficult to do this without leaving a trace. The situation is similar to modern forensic science, where the smallest trace of organic material can be used for DNA fingerprint-ing. In the case of data analysis, natural mechanisms generate data having patterns that are different from those generated by human thought. A simple illustration of this can be found by asking someone to write down a series of 'random' digits and then subjecting the series to statistical tests to see if the numbers are in fact random. People typically depart from randomness in a number of ways. For example, the deviation from roughly equal proportions of digits is often smaller than one would expect by chance.

I know of one case where doubt was cast on the recorded values of pulse rates in a medical study. The standard approach to taking pulse rates is to count the number of beats which occur in fifteen seconds, and multiply by four to give the rate per minute. Unfortunately, in one data set, none of the figures were divisible by four. When confronted, the research nurse admitted to having made up the numbers.

More generally, data analytic techniques can be used to detect when some odd or unusual process has generated the data – not because the data are simply made up, but because they arise from some process other than that claimed. For example, I encountered a case in which questions had arisen about the process by which a particular pharma-ceutical product had been manufactured. Company A had patented a manufacturing process, and claimed that company B was using this process illegally. All drug manufacturing processes leave a fingerprint of trace impurities, which can be detected by subtle chemical analysis. This fingerprint can then be subjected to a data analysis which will tell how likely it is that the material was produced by any given process.

In a completely different area, the ideas can also be used to detect student cheating in tests, coursework, and examinations. Cheating can take various forms. One important form is plagiarism, in which material is copied from the web or other sources, without attribu-tion. Software tools have been developed which trawl the web and match coursework essays against web documents, and some of these tools have been highly successful in detecting such dishonesty.

Of course, plagiarism is not restricted to students. It can even happen at the highest levels of government, and has been used in support of wars. In February 2003, the British Government published a dossier on Iraq's intelligence organizations. This document was also cited by Colin Powell in a speech that month. And where did the material in this document originate? Here is an extract from an article by Glen Rangwala, a Cambridge University politics lecturer: (from http://www.casi.org.uk/discuss/2003/msg00457.html)

> *In preparation for Powell's presentation at 15:30 GMT today, I had a look at the third British government's 'dossier' released last Thursday, "Iraq – Its Infrastructure Of Concealment, Deception And Intimidation" (30 January 2003).*
>
> *...*
>
> *The document claims to draw 'upon a number of sources, including intelligence material' (p. 1, first sentence).*
>
> *Now this is a bit misleading.*
>
> *More precisely, the bulk of the 19-page document (pp. 6–16) is directly copied without acknowledgement from an article in last September's Middle East Review of International Affairs entitled 'Iraq's Security and Intelligence Network: A Guide and Analysis'.*
>
> *...*
>
> *The author of the piece is Ibrahim al-Marashi, a postgraduate student at the Monterey Institute of International Studies. He has confirmed to me that his permission was not sought; in fact, he didn't even know about the British document until I mentioned it to him.*
>
> *It's quite striking that even Marashi's typographical errors and anomolous [sic] uses of grammar are incorporated into the Downing Street document. For example, on p. 13, the British dossier incorporates a misplaced comma:*
>
> *'Saddam appointed, Sabir 'Abd al-'Aziz al-Duri as head'..*
>
> *Likewise, Marashi's piece also states:*
>
> *'Saddam appointed, Sabir 'Abd al-'Aziz al-Duri as head'..*
>
> *The other sources that are extensively plagiarised in the document are two authors from Jane's Intelligence Review:*
>
> *Ken Gause (an international security analyst from Alexandria, Virginia), 'Can the Iraqi Security Apparatus save Saddam' (November 2002), pp. 8–13.*
>
> *Sean Boyne, 'Inside Iraq's Security Network', in 2 parts during 1997.*

None of the sources are acknowledged, leading the reader to believe that the information is a result of direct investigative work, rather than simply copied from pre-existing internet sources.

Another form of student cheating involves copying from each other, or colluding in answering questions. Statisticians Niall Adams, Nick Heard and I developed a software tool for detecting such cases in statistics coursework. The work took the form of a dozen or so extended mathematics questions, which the students were supposed to work on independently in their own time. We coded their answers based on syntactic descriptions – like the copied misplaced comma in the Iraq dossier example, but for us in terms of how often they used various mathematical symbols – and then our computer program calculated measures of how similar their answers were. Based on the background variation between students (most of whom we assumed were honest), we were able to calculate the probability that certain very similar scripts could have been that similar if the students had not worked together. A very low probability rouses one's suspicions.

Plagiarism and cheating are examples of fraud, and fraud is a constant problem in commercial situations. I have made a particular study of fraud in banking contexts. It can take many forms – and the situation has been aggravated by advancing technology that, while it makes life easier in so many ways, also opens up the scope for new kinds of fraud. Of course, there are many tools for preventing fraud in the first place: chip and PIN systems with credit cards, fluorescent fibres, watermarks and holographs in banknotes, passwords on computer systems and bank accounts, and so on. Despite this, once a fraud has been perpetrated, data analysis is needed to detect it. Obviously, in banking applications, the sooner one detects a fraud the better: it would be nice to know that your credit card had been cloned immediately the thief started to use it, and not three months later.

It is convenient to divide banking fraud detection tools into two main classes. One class is based on data describing previous frauds: such tools then measure how similar are current transaction patterns to known fraudulent patterns. For example, sudden use of credit cards to purchase many small electrical items can be a sign of fraud (such items are easily sold on) – though of course, it need not: perhaps the card owner is buying such items as Christmas presents. The other class is based on detecting anomalous behaviour. Data from past transactions shows how a customer typically behaves, and if their behaviour suddenly departs from this then it should arouse suspicion. For example, if someone who regularly takes out $200 from an ATM every Friday night suddenly, over one weekend, withdraws several successive $500 sums, then one might reasonably be suspicious. A third, but relatively minor class of fraud detection tools is based on empirical properties for how data behave. An example is the use of the Benford distribution, described in chapter 6, for detecting fabricated numbers in accounts. This distribution tells us how often the first digit in a set of values should be 1, how often it should be 2, and so on. Departure from this empirically verified and theoretically justified distribution might arouse one's suspicions, at least for certain kinds of data.

One of the difficulties in banking fraud detection is that most transactions are legitimate – a rate of one fraudulent transaction every thousand transactions is the ball-park figure, though obviously it depends on many factors. Now, a detection method which correctly identifies ninety-nine per cent of the legitimate records as legitimate and ninety-nine per cent of the fraudulent records as fraudulent might be regarded as a highly effective system. But now consider: on average, in every 100,000 transactions, one in a thousand, that is 100, will be fraudulent and 99,900 will be legitimate. Our 'highly effective' system will correctly classify 99 of the 100 fraudulent transactions as fraudulent. But it will also classify 1 in 100 of the 99,900 legitimate transactions as fraudulent. That is 999 legitimate transactions will be mistakenly classified as fraudulent. Altogether, then, the system has classified 999 + 99 = 1098 transactions as fraudulent – *but only ninety-nine of these, that is nine per cent, are really fraudulent*. Now think of the cost of investigating all 1098 closely – both in terms of financial cost and the potential damage to customer relations.

Money laundering is another area of financial fraud where data analysis plays a crucial role. Once again, the basic tools are methods for detecting unusual patterns of transactions, or transactions following patterns which are known to be associated with money laundering. Sometimes the detection of such patterns leads to changes in the law. In 1970, for example, the United States Bank Secrecy Act required that banks report all currency transactions of over $10,000 to the authorities. Of course, money launderers adapt to any new constraint, and the natural response to this law was to split their money transfers into multiple transactions of just less than $10,000 a time. This, in turn, led to tools which searched the data for multiple such transactions. And so it goes on: the battle against fraud is an evolutionary one, involving a leapfrogging of fraudsters and the authorities.

Detection of much commercial fraud, and banking fraud is a good example, is made tougher by an (understandable) unwillingness of banks to admit that it takes place. It hardly inspires confidence amongst potential customers to learn that the bank lost $100 million dollars to fraud last year, for example. This explains why a senior banker told me, at a recent conference on banking crime, 'we don't have any fraud'. One might also wonder if the fraud which is detected is merely the tip of the iceberg. Since there are many cases of fraud not being discovered until long after the event, or on the tenth or twentieth time the fraudster has done it, presumably at least some fraud goes undetected. The fraudster who decided that enough was enough on their ninth job, for example.

Banking data sets are often large, and another area involving large data sets is telecommunications. Telecommunications networks can generate several gigabytes of data per day, so the data analytic challenges are considerable. There are various estimates of the cost of telecommunications fraud, but some go as high as $10 billion per year. The figure depends on what is included and how it is estimated, on whether, for example, it includes hard currency fraud (real money paid by someone for a good or service, which is stolen – such as by telephone or Internet credit card fraud, calls to numbers that charge for access etc.) as well as soft currency fraud (for example, fraudulently obtaining a subscription to a service, with no intention of paying).

Once again, analysis of the phone records will reveal such fraud, though, also once again, detection is more useful the earlier it takes place. Signatures of typical behaviour patterns can be used to detect the changes of behaviour associated with fraudulent use, though one must recognize that behaviour patterns change over time – even within a day.

An area related to telecoms fraud is computer intrusion or hacking. On Thursday, 21 September 2000, a sixteen-year-old boy was jailed for hacking into both the Pentagon and NASA computer systems. The key data analytic tool for detecting hacking is sequence analysis: recognizing something anomalous about the sequence of commands being used to access or manipulate a system.

Of course, not all forensic statistics is either recent or concerned with high-tech applications. Recall the discovery made by Ridolfo Livi at the end of the nineteenth century, described in chapter 6. He found that the existence of two distinct peaks in a data set, which had been attributed to the existence of two different racial groups in the population, was purely the result of how numbers, originally recorded to the nearest centimetre, had grouped together when translated into inches and rounded to the nearest inch.

Epilogue
The beginning

While you can't have your cake and eat it,
You can sell your data and keep it.
David J. Hand

The world has experienced dramatic changes over the past few centuries. In large part, those changes are a consequence of representing things in terms of numbers, in terms of data. Data form the basis of record systems that tell us how economic, social, or physical systems have behaved in the past. They form the basis of models that can be manipulated, tested and explored to yield understanding of the world. Such models enable us to predict the future, to gather information about the likely potential impact of different actions, so that we can make intelligent choices. Data, in fact, form the very basis of modern civilization. The old and familiar aphorism says that 'knowledge is power', but that knowledge is built from data, and it is that knowledge, constructed from the basic building blocks of data, which has enabled our civilization to progress to its current state.

Especially in recent decades, we have witnessed extraordinarily dramatic developments in the sciences and technologies of data. Established scientific disciplines have been revolutionized by advances in data capture, manipulation and analysis, and entirely new scientific fields have sprung up driven by the same developments. Indeed, it is difficult to imagine any area of modern science which does not hinge on advanced tools for data manipulation and analysis. Furthermore, new industries have appeared that are built on data, industries such as survey sampling, geodemographics, customer relationship management and, that modern revolution in information technology, the Internet. Indeed, no modern corporation, whatever the area in which it functions, could exist today without highly advanced data manipulation systems (at the very least, the complexity of modern taxation and employment legislation sees to that!). All of these advances mean that the world we wake up to each morning is very different from the world that our ancestors woke up to.

There is always a tendency, when looking back in this way, to imagine that one has reached the end of the story, that we are at the pinnacle. As far as the story of data goes, however, that would clearly be a mistake. The developments described above are merely the beginning: we are simply standing on a gentle slope at the foot of the data mountain. The fact is that the pace of change is accelerating. What seem to us like huge data sets now will soon become commonplace. Even larger ones, constituted from streams of data arriving at even more awesome rates, will become everyday. The cost of data will fall, as increasingly advanced measuring instruments and electronic data capture technologies continue to develop. Moore's Law and its variants suggest that our data storage capacities and our ability to manipulate the data will continue to grow. Perhaps quantum computing will provide a revolution in computational infrastructure enabling the development of data analytic tools beyond those we can even imagine. Surreptitious observation of people, either directly via webcams and CCTV, or indirectly via monitoring of transactions, interactions with others, movement and transport, and in a host of other ways will continue to increase.

All of these developments will open up other, wider, social and cultural questions. In particular, any advanced technology can be used for good or ill. Technology itself is neutral: it is those who wield that technology who carry the moral and ethical burden of responsible use. The Internet is a prime example of a data-driven technology that has, in the space of just a few years, revolutionized our lives and that has already presented challenging questions. For example, questions have been raised about the accuracy of the information it carries (e.g., the case of Wikipedia, which allowed anyone to create Internet encyclopedia entries), about its potential for invading privacy (e.g., the recent case of an anonymous sperm donor being identified after just two web searches), and about its capacity to criticize unfairly (e.g., the controversy surrounding the RateMyProfessors.com website). And there are deeper questions hinging around the fact that the Internet deals purely in information, at the bottom level, in data. The economics of data are different from the economics of manufacturing industry, and even of service industries. Data can be copied, at negligible cost, as many times as one likes. This is very different from what happens when we build a house, or wash a car.

The baseball philosopher Yogi Berra said 'prediction is a risky business, especially when it's about the future'. But of one thing we can be sure: we will depend more and more on the collection, manipulation and analysis of data. Our future will depend on it.

Further Reading

Alder K. 2002. *The Measure of All Things: The Seven-Year Odyssey That Transformed the World*. London: Little, Brown.

Babbage C. 2004. *Reflections on the Decline of Science in England and on Some of its Causes*. Whitefish, Montana: Kessinger Publishing Co.

Bacon F. 1620 (1994). *The Novum Organum*. Transl. P. Urbach and J. Gibson. Chicago, Illinois: Open Court Publishing Co.

Bacon F. 2002. *Francis Bacon: the Major Works*. Oxford: Oxford University Press.

Bennett J., Cooper M., Hunter M. and Jardine L. 2003. *London's Leonardo: the Life and Work of Robert Hooke*. Oxford: Oxford University Press.

Bentham J. 1995. *The Panopticon Writings*, pp. 29–95. ed. M. Bozovic. London: Verso.

Bernal J. D. 1954. *Science in History*. London: Watts and Co.

Berry M. and Linoff G. 2000. *Mastering Data Mining*. New York: John Wiley & Sons.

Boring E. G. 1929. *A History of Experimental Psychology*. New York: Century.

Boyle D. 2000. *The Tyranny of Numbers: Why Counting Can't Make us Happy*. London: Harper Collins.

Clarke A. C. 1962. *Profiles of the Future*. London: Gollancz.

Cook A. 1994. *The Observational Foundations of Physics*. Cambridge: Cambridge University Press.

Copernicus N. 1995. *On the Revolutions of the Heavenly Spheres*. Amherst, NY: Prometheus Books.

Courtney L. H. 1895. To my fellow-disciples at Saratoga Springs. *The National Review*, **26**, 21–26.

Crosby A. 1997. *The Measure of Reality: Quantification and Western Society, 1250–1600*. Cambridge: Cambridge University Press.

Dawes R. M. and Smith T. L. 1985. Attitude and opinion measurement. In *The Handbook of Social Psychology*, Volume I, 3rd edition, eds G. Lindzey and E. Aronson, New York: Random House, pp. 509–566.

Del Rio, M. 2000. *Investigations into Magic*. ed. and transl. P. G. Maxwell, Manchester: Manchester University Press.

Desrosières A. 1998. *The Politics of Large Numbers: A History of Statistical Reasoning*. Cambridge, Massachusetts: Harvard University Press.

Franklin B. 1963. Pennsylvania Assembly: Reply to the Governor, November 11th. In *The Papers of Benjamin Franklin*, ed. Leonard W. Labaree, vol. 6, p. 242.

Galileo Galilei 1615. *Letter to the Grand Duchess*. http://www.galilean-library.org/christina.html.

Galton F. 1863. *Meteorographica: Methods of Mapping the Weather*. London: Macmillan. Available in electronic format at www.galton.org.

Gertner J. 2004. The very, very personal is the political. *New York Times*, 15 February.

Gilbert W. 1600. *De Magnete*. London: Chiswick Press.

Goodhart G. 1984. *Monetary Theory and Practice, the UK Experience*, London: Macmillan Press.

Hall M. P. 1990. *Paracelsus, His Mystical and Medical Philosophy*. Philosophical Research Society Inc., U.S.

Hamilton A. 1787. The consequences of hostilities between the States from the New York Packet. *Federalist Paper No. 8*. 20 November.

Hand D. J. 2004. *Measurement Theory and Practice: The World Through Quantification*. London: Arnold.

Hand D. J., Mannila H. and Smyth P. 2001. *Principles of Data Mining*. Cambridge, Massachusetts: MIT Press.

Hargittai I. 2002. *The Road to Stockholm: Nobel Prizes, Science, and Scientists*. Oxford: Oxford University Press.

Hastie T., Tibshirani R. and Friedman J. 2001. *The Elements of Statistical Learning: Data Mining, Inference, and Prediction*. New York: Springer.

Heinlein R. A. 1973. *Time Enough for Love*. New York: Ace.

Huff D. 1954. *How to Lie with Statistics*. London: Gollancz.

Hume D. 1999. *An Enquiry Concerning Human Understanding*. Oxford: Oxford University Press.

Huygens C. 1920. De Ratiociniis in Ludo Alae. In *Oeuvres Complètes de Christiaan Huygens, vol. XIV*. The Hague: Martinus Nijhoff.

Ifrah G. 1998. *The Universal History of Numbers: From Prehistory to the Invention of the Computer*. London:The Harvill Press.

Inwood S. 2002. *The Man Who Knew Too Much*. London: Macmillan.

Jardine L. 2003. *The Curious Life of Robert Hooke: The Man Who Measured London*. NewYork: HarperCollins.

Jones R. F. 1961. *Ancients and Moderns: A Study of the Rise of the Scientific Movement in Seventeenth-century England*. NewYork: Dover.

Judson H. F. 2004. *The Great Betrayal: Fraud in Science*. Orlando: Harcourt Inc.

Kepler J. 1992. *New Astronomy*. Transl. W. H. Donahue. Cambridge: Cambridge University Press.

Kepler J. 1997. *The Harmony of theWorld*. Transl E. J. Aiton, A. M. Duncan and J.V. Field. Philadelphia, PA: American Philosophical Society.

Keynes J. M. 1947. Newton the Man. In *Newton Tercentenary Celebrations 15–19 July 1946*, pp. 27–34. Cambridge: Cambridge University Press.

Leighton G. and McKinlay P. L. 1930. *Milk Consumption and the Growth of School Children*. Edinburgh: HMSO.

Littlewood J. E. 1953. *A Mathematician's Miscellany*. London: Methuen.

McDowell I. and Newell C. 1996. *Measuring Health: A Guide to Rating Scales and Questionnaires*. NewYork: Oxford University Press.

McGreevy T. 1995. *The Basis of Measurement: 1. Historical Aspects*. ed. P. Cunningham. Chippenham: Picton Publishing.

Meehl P. E. 1986. Causes and effects of my disturbing little book. *Journal of Personality Assessment*, **50**, 370–375.

Minow N. N. *et al.* 2004. *Safeguarding Privacy in the Fight Against Terrorism: Report of the Technology and Privacy Advisory Committee*. Washington, DC, US Department of Defense, March.

Moore G. E. 1965. Cramming more components onto integrated circuits. *Electronics*, **38**, 8, 114–117.

NASA 2000. *Probabilistic risk assessment: a bibliography*. NASA/SP-2000-6112. NASA Scientific and Technical Information Program.

Norton Wise M. 1995. ed. *The Values of Precision*, p. 358. Princeton, NJ: Princeton University Press.

Payne J. F. 1900. *T. Sydenham*. London:T. F. Unwin.

Pears Cyclopedia 1990. 99th edition, pp. G27, G31. London: Penguin.

Porter H. W. 1853–54. On some points connected with the education of an actuary. *Assurance Magazine*, **4**, 108–118.

Roberts J. M. Jr and Brewer D. D. 2001. Measures and tests of heaping in discrete quantitative distributions. *Journal of Applied Statistics*, **28**, 887–896.

Slobogin C. 2002. Public privacy: camera surveillance of public places and the right to anonymity. *Mississippi Law Journal*, **72**, 213–315.

Spedding J., Ellis R. L. and Heath D. D. eds 1879–1890. *Works of Francis Bacon*, London.

Sprat T. 1959. *History of the Royal Society*. ed. by J. I. Cope and H. W. Jones. St. Louis: Washington University Press.

Stamp J. 1929. *Some Economic Factors in Modern Life*. London: P. S. King.

Stigler G. I. 1949. *Five Lectures on Economic Problems*. Freeport, NY: Books for Libraries Press.

Stigler S. M. 1986. *The History of Statistics: The Measurement of Uncertainty Before 1900*. Cambridge, Massachusetts: Harvard University Press.

Strathern M. 1997. Improving ratings: audit in the British university system, *European Review*, **5**, 305–321.

Student 1931. The Lanarkshire milk experiment. *Biometrika*, **23**, 398–406.

Turnbull H. (ed.) 1959. *The Correspondence of Isaac Newton: Vol. 1.* Cambridge: Cambridge University Press.

Washington G. 1787. *Letter from the Federal Convention President to the President of Congress, Transmitting the Constitution*. 17 September.

Westman R. S. 1980. The astronomer's role in the sixteenth century, a preliminary survey. *History of Science*, **18**, 105–147.

Willenborg L. and de Waal T. 2001. *Elements of Statistical Disclosure Control*. New York: Springer.

Index

INDEX

CL

303.
483
3
HAN

5001023886